FM 4-01.502 (FM 55-502)

I0171262

Army Watercraft Safety

May 2008

Headquarters, Department of the Army

Field Manual
No. 4-01.502 (55-502)

Headquarters
Department of the Army
Washington, 1 May 2008

Army Watercraft Safety

Contents

Page

Distribution Restriction: Approved for public release; distribution is unlimited.

*This publication supercedes FM 55-502, 23 December 1996.

Tables

Preface

This manual provides guidance and technical information relevant to safety and survival equipment/systems used by US Army watercraft. Items listed in Appendix A are provided by the Marine Safety Office as safety equipment recommended for use aboard Army Watercraft. Vessel Technical Manuals and Basic Issue Items lists should be consulted for required equipment.

The manual contains guidance, instructions, technical data, illustrations, and procedures pertinent to the application, inspection, modification, maintenance, and the use of safety equipment, safety policies, and survival systems. The primary users of this manual are watercraft masters and key personnel engaged in the supervision, operation, or maintenance of US Army watercraft.

This publication applies to the Active Army, the Army National Guard/Army National Guard of the United States, and the United Sates Army Reserve unless otherwise stated

The proponent of this publication is the United States Army Training and Doctrine Command (TRADOC). The preparing agency is the United States Army Transportation Center and School. Submit written comments and recommendations for improving this publication on DA Form 2028 (Recommended Changes to Publications and Blank Forms) to Commander, USATC&FE, ATTN: ATZF-OCT-S (Marine Safety Office), Fort Eustis, VA 23604-5407. Send comments and recommendations on electronic DA Form 2028 by email to marinesafety@conus.army.mil and to LeeeATCLFC@conus.army.mil (Attn: Doctrine Division).

Unless this publication states otherwise, masculine nouns and pronouns do not refer exclusively to men.

Introduction

PURPOSE

This manual provides a single source of guidance and technical information relevant to rescue equipment, survival systems and safety policies used on Army watercraft by Army watercraft personnel.

NOTE: Deviations from configurations of equipment presented in this manual are not recommended and are not authorized. To maintain standardization and to preclude the dangers of operating with potentially unsafe equipment, modifications to rescue equipment and survival systems are not authorized.

ADMINISTRATION

Because watercraft units have a unique requirement for the specific vessel Annual Safety Survey, commanders will appoint, in writing, a person to manage the unit's Annual Safety Survey program in addition to the generic duties of the Unit Safety Officer. This individual is responsible for the unit administration and coordination of the safety equipment inspection and maintenance requirements outlined in this manual.

CONTENTS

This manual contains guidance, instructions, technical data, drawings, illustrations, procedures, and descriptions pertinent to the configuration, modification, application, inspection, fabrication, maintenance and repair, and the use of rescue equipment and survival systems. The Transportation Branch Marine Safety Office, Fort Eustis, VA, will formulate, produce, and distribute applicable changes to the equipment and to respective chapters of this manual.

Warnings, Cautions and Notes The following definitions apply to "WARNINGS," "CAUTIONS," and "NOTES" found throughout this manual.

WARNING

Operating or maintenance procedures, techniques may result in personal injury or loss of life if not carefully followed.

CAUTION

Operating maintenance procedures, techniques, may result in damage to equipment if instructions are not carefully followed.

NOTE: Operating or maintenance procedures, techniques, and so on that are considered essential to emphasize.

USE OF THE WORDS "SHALL," "WILL," "MAY," AND "SHOULD"

The following definitions apply to the words "shall, will, may," and "should" throughout this manual.

- Use of "shall" and "will" indicates a mandatory requirement.
- Use of "may" and "should" indicates an acceptable or suggested means of accomplishment.

UPDATING

This manual will be updated periodically by the issuing of a List of Effective Pages that includes all original, revised, added, and deleted pages. This list will be inserted in the front of each volume, immediately following the title page. Revised and added pages, appropriately dated, will be issued with the change and will be inserted into this volume according to page numbers. The replaced and/or deleted pages and the superseded List of Effective Pages will be discarded. National Stock Numbers (NSN) and associated equipment listed may change as technologies advance. As recommended, interim changes will be submitted for insertion to this manual by means of the US Army Watercraft Safety Advisory from the Marine Safety Office. All directives shall be complied with that are issued after the date of the latest change. These directives will be incorporated into the next change or revision of this manual.

GENERAL

Regulatory guidance for rescue and survival equipment and systems are contained in AR 56-9. Policies and procedures pertaining to all other aspects of the equipment and systems are stated in this manual.

ALLOWANCE

Allowances of rescue and survival equipment are determined by the type of vessel and mission. Additional equipment may be required to provide for passengers and maintenance. Specific allowances may be found in the appropriate watercraft Basic Issue Items (BII) list. Appendix A of this manual provides a quick reference guide, but must be verified with the BII.

Chapter 1

Marine Safety Program

The Army's watercraft safety program is designed to help mariners perform their mission in a safe manner and to give the Risk Level Decision Maker the necessary support/advise tools to make an informed decision. The program uses vessel safety surveys, a hazard tracking system, and institutional training to ensure that watercraft and the personnel assigned are able to maintain operational readiness and mission effectiveness. Topics covered in this section are:

- Safety Surveys and inspections
- Army Accident Reporting and Investigation
- Watercraft Safety Notification and Readiness
- Safety Inspections

SAFETY SURVEYS AND INSPECTIONS

1-1. **Surveys.** The Transportation Branch Marine Safety Office is assigned the task of performing Marine Safety Surveys to meet the Department of the Army's triennial requirement. All Army watercraft will undergo a safety survey every three (3) years conducted by the Transportation Branch Safety Office (TBSO), Marine Safety Specialist (MSS), Fort Eustis, VA., and annually by the Battalion or below. Surveys will not be conducted on watercraft in overhaul, at sea, or within the first three months after being placed in service. The purpose for the safety survey is to:

- Uphold and maintain the safety posture of Army watercraft as related to readiness.
- Provide compliance with Army and other federal safety regulations.
- Assess the level of safety standardization within the Army watercraft field.
- Provide on–site assistance for crew safety training.
- Accumulate lessons learned from Army watercraft crews.

1-2. **Survey Guides.** Safety Survey Guides are designed to be generic to vessel classes and are periodically updated as both vessel safety equipment and governing laws/regulations improve. Standardized Safety Guides have been developed and are available from the Transportation Branch Marine Safety Office and the Army Electronic Product Support web site.

1-3. **Inspections.** Safety inspections are the responsibility of the Commander. Army watercraft must follow Army Regulations, Department of Defense Regulations, United States Public Laws and International Maritime Laws. Both Coastal and International laws allow the Department of the Army to establish different criteria as long as these criteria do not adversely impact any other Maritime organization or Maritime operation. The vessel master is ultimately responsible for insuring vessel inspections and surveys occur as required for his or her assigned vessel, with special attention to rescue and survival equipment.

1-4. Both the public laws and DOD/DA Regulations require Safety inspections of each watercraft to be performed within specific time frames. Public law requires annual inspections be performed by the owning organization and periodic inspections by law enforcement agencies of the national flag of the vessel. DA Regulations require an annual safety inspection of all equipment and facilities managed/owned by each specific command. Additionally, DOD Regulations require a triennial safety inspection of all vessel equipment and management processes be conducted by a major component command level. The Army has assigned this task to Commander, Training and Doctrine Command, who further assigned the task to the Chief of Transportation.

ARMY ACCIDENT REPORTING AND INVESTIGATION

DEFINITIONS

1-5. **Accidents.** An Army accident is defined as an unplanned event or series of events. For watercraft, these events could result in one or more of the following:

- Accidents occurring while loading, off-loading, or receiving services dockside.
- Damage to Army property (including government-furnished material, government property, or government-furnished equipment provided to a contractor).
- Accidents occurring during amphibious or on-shore warfare training operations.
- Injury (fatal or nonfatal) to on or off duty military personnel.
- Injury (fatal or nonfatal) to on-duty Army civilian personnel. Includes non-appropriated fund SOLDIERS and foreign nationals employed by the Army when incurred during performance of duties while in a work-compensable status.
- Occupational injury or illness (fatal or nonfatal) to Army military personnel, Army civilian SOLDIERS, non-appropriated fund SOLDIERS, or foreign nationals employed by the Army.
- Injury or illness (fatal or nonfatal) to non-Army personnel or damage to non-Army property.

1-6. **Report.** A report is defined as the initial notification process--informing the chain of command and other required parties in the time frames prescribed that an Army defined accident has occurred. Appendix B has the telephonic notification DA Form 7306 (*Worksheet for Telephonic Notification of Ground Accident*), used for the initial notification.

1-7. **Record.** Recording is the actual completion of the required form(s) (DA Form 285 [*U.S. Army Accident Report*] or DA Form 285-AB-R [*U.S. Army Abbreviated Ground Accident Report*]), part of the appointed Accident Investigating Officer's requirements.

1-8. The types of watercraft under the jurisdiction of the DA are those that are:

- Operated, controlled or directed by the Army. This includes watercraft furnished by a contractor or another Government agency when operated by Army watercraft personnel.
- Loaned or leased to non-Army organizations for modification, maintenance, repair, test, and contractor, research, or development projects for the Army.
- Under test by Army agencies responsible for research, development, and test of equipment.
- Under operational control of a contractor for the Army.

REPORTING IMPORTANCE

1-9. Army accidents are reported and investigated to identify problem areas (deficiencies) as early as possible in order to save personnel and equipment. Changes, corrections, and countermeasures can be developed and implemented to these deficiencies before more people are hurt or killed or equipment is damaged, destroyed, or lost. If an accident is never reported, the local command and required Department of the Army Agencies will not know there is a problem. Unreported accidents lead to repeat occurrences. Errors must be addressed to prevent future loss of man-hours, life and equipment. Reporting and investigating Army accidents in a complete and timely manner is an extremely important function. Procedures for watercraft accident investigation and written reports are outlined in AR 56-9 and DA Pamphlet 385-40.

MASTER/OPERATOR REPORTING RESPONSIBILITIES

1-10. If an Army watercraft is involved in an accident, the master/operator must report the accident (initial notification) via any electronic means available, within 24 hours of the occurrence to their local command, commands of concern, and the Transportation Branch Marine Safety Office in accordance with AR 385-10 and DA PAM 385-40. If not previously reported by the vessel master / operator, the local command has the responsibility, per AR 385-10; DA Pam 385-40 and AR 56-9, to pass on (notify) these reported occurrences to the Transportation Branch Marine Safety Office at marinesafety@conus.army.mil and the Commander, USACRC. The primary method for immediate notification is through the Web–based initial notification (IN) tool located on the USACRC Web site at https://crc.army.mil/home. Checklists for compiling the necessary ground or aviation accident information to complete the IN tool report are provided online. The secondary method for immediate notification is by telephone (DSN 558–2660/558–3410, commercial (334) 255–2660/255–3410). At a minimum, notification will include the information on DA Form 7306. See Appendix B for an example form. Reporting by phone or using the IN tool meets the requirement for reporting within 24 hours. The master/operator does not report using the DA Form 285 report. This chapter does not negate the master's responsibility to report any applicable watercraft accident, injury, or death involving commercial watercraft or property to the US Coast Guard; or the US Coast Guard's responsibility to investigate a watercraft-related accident.

ACCIDENT REPORTS

1-11. Army investigated accidents must be recorded on the appropriate forms. Currently this is DA Form 285. The recorded report is prepared by the appointed Investigating Officer according to DA Pamphlet 385-40 requirements. Only a commander in the chain of command can appoint the Accident Investigating Officer. This report is intended only for accident prevention purposes and will not be used for administrative or disciplinary actions within the DOD.

NOTE: To inform the Investigating Officer of marine-related information pertinent to the investigation, the following additional information will be included in DA Form 285 as an enclosure:

- Time and place of commencement of voyage and destination.
- Current (direction and force).
- Wind (direction and force).
- Visibility in yards.
- Tide and sea conditions.
- Name of person in charge of navigation and persons on the bridge.
- Name and rank of lookout and where stationed.
- Time when bridge personnel and lookouts were posted on duty.
- Course and speed of watercraft.
- Number of passengers and crew on board.
- Names of crew and passengers.
- Copies of all pertinent log entries.
- List of the names and addresses of witnesses.
- When steering gear and controls were last tested.
- When and where compasses were last adjusted and the magnetic deviation, if any, at time of accident.
- Statement of any outside assistance received.
- Diagrams of damage and pertinent documents.
- Photos of damage.
- Any further details not covered above.

1-12. **Recordable (Investigated) Accidents.** It is mandatory that all Army accidents are reported, regardless of class, to the local activity or installation safety office. However, only recordable (investigated) accidents require completion and submission of DA Form 285. These recordable accidents include Classes A, B, and Aviation Class C accidents (see AR 385-10 for details). The Army classifies recordable accidents by severity of injury and property damage. These classes (A through D) are used to determine the appropriate investigative and reporting procedures, and are described below.

- **Class A.** The total cost of reportable damage is $1,000,000 or more. An Army aircraft, watercraft, missile, or spacecraft is destroyed, missing or abandoned; or an injury / occupational illness results in a fatality or permanent total disability. Unmanned Aircraft Systems (UAS) reporting is based on repair or replacement cost. Loss of a UAS is not a Class A Accident unless more than $1,000,000 to repair or replace.

- **Class B.** The total cost of reportable property damage is $200,000 or more, but less than $1,000,000. An injury and/or occupational illness results in permanent partial disability; or three (3) or more people are hospitalized as inpatients as the result of a single occurrence.
- **Class C.** The total cost of property damage is $20,000 or more, but less than $200,000. A nonfatal injury or occupational illness that causes one (1) or more days away from work or training beyond the day or shift on which it occurred; or a disability at any time (that does not meet the definition of Class A or B and is a lost time case).
- **Class D.** The cost of property damage is $2,000 or more, but less than $20,000; a nonfatal injury or illness resulting in restricted work, transfer to another job, medical treatment greater than first aid, needle stick injuries and cuts from sharps that are contaminated with another person's blood or other potentially infectious material, medical removal under medical surveillance requirements of an OSHA standard, occupational hearing loss, or a work-related tuberculosis case.

NOTE: Property damage is defined as the cost to repair or replace to original condition. Property damage costs are separated from personnel injury/illness costs for classifying A through C accidents.

1-13. For Army accidents that require a DA Form 285, the commander having general court-martial jurisdiction over the installation or unit responsible for the operation, personnel or materiel involved in the accident will make sure of the following:

- Appoint an Investigating Officer and any support technical assets required.
- An investigation is performed to obtain the facts and circumstances of the accident for accident prevention purposes only (DA Pam 385-40).
- Collateral investigations are used to obtain and preserve all available evidence for use in litigation, claims, disciplinary action, or adverse administrative actions. They are essential for the protection of the privileges afforded to accident investigation reports, as they ensure there is an alternative source of evidence for use in legal and administrative proceedings. Although non-privileged information acquired by a safety accident investigator shall be made available to the collateral investigation, the latter is conducted independently and apart from other types of accident investigations.
- Evidence is preserved in accordance with DA Pam 385-40.
- Personnel interviewed during an Army Accident Investigation do not fill out or sign any Sworn Statements. All statements are recorded in the Investigating Officer's handwriting and the third person.
- DA Form 285 is completed according to instructions on the form and DA Pamphlet 385-40. The form must be forwarded through the installation safety office to the Army Safety Center and the Transportation Branch Marine Safety Office for recording in the Army Safety Management Information System (ASMIS) within 30 days of the accident. Army Reserve reports are sent to the Army Reserve Safety Directorate at this address: Commander, US Army Reserve Command, ATTN: ARRC-SA, 1401 Deshler Street SW, Fort McPherson, GA 30330.
- See DA Pamphlet 385-40 for accidents that require a board investigation.

1-14. **Collateral Investigation Reports.** A collateral investigation report is required in many cases for class A, B, or C accidents to make and preserve a record of the facts for litigation, claims, and disciplinary and administrative actions. These investigations are conducted in accordance with AR 15-6 and not the procedures in DA Pamphlet 385-40. A collateral investigation is required on all fatal accidents. It is also required for those accidents that generate a high degree of public interest or are likely to result in litigation for or against the Government. Releasing or sharing information gathered under accident investigation must be strictly controlled IAW DA Pam 385-40. The appointed Accident Investigation Officer cannot be appointed as a collateral investigation officer on the same accident.

WATERCRAFT SAFETY NOTIFICATION AND READINESS

SAFETY MESSAGES

1-15. Safety messages are used to transmit information concerning hazards identified in equipment that poses a danger to personnel or the system itself. The following list identifies the types of messages that the user will encounter and their meanings.

- **Emergency message.** Orders to cease operation/use of a specific model, series, or design of equipment failure to adhere to the message will have catastrophic results to the system.
- **Ground Precautionary action Message (GPM).** Electronically transmitted message pertaining to any defect or hazardous condition, actual or potential, where a medium or low-risk safety condition has been determined per AR 385–16 that can cause injury to Army personnel or damage to Army equipment.
- **Maintenance Action Message (MAM).** Electronically transmitted messages that convey equipment maintenance, technical or general interest information.
- **Maintenance information message.** Electronically transmitted messages that convey general interest information that is permissive in nature.
- **Safety of Use Message (SOUM).** Message pertaining to any unsafe condition, determined per DA Pam 385–16. May be of an operational, technical, one-time inspection, or advisory nature.

WEB SITE INFORMATION (HTTP://AEPS.RIA.ARMY.MIL)

STANDARDIZED PRODUCTS

1-16. These products are acquired through acquisition support and are found in Appendix A. The field maintains standardized equipment for the mariner to use. This is found in the Basic Issue Items Listing (BII) onboard the vessels and must be inspected at specified intervals. The inspections are used to identify equipment for serviceability; any items found to be unserviceable or obsolete will be removed from the inventory immediately. These items include, but are not limited to life saving equipment and emergency equipment.

TEST DRILLS AND INSPECTIONS (TDI)

Emergency Drills

1-17. Drills must be held in accordance with AR 56-9. Additionally, drills must be held before sailing when a vessel enters service for the first time, after modification of a major character, or when a new crew is engaged.

1-18. Abandon-ship drills must include summoning persons on board to muster stations with the general alarm followed by drill announcements on the public address or other communication system and ensuring that the persons on board are made aware of the order to abandon ship, reporting to stations and preparing for the duties described in the muster list, checking that persons on board are suitably dressed, checking that lifejackets or immersion suits are correctly donned.

1-19. Each fire drill must include reporting to stations and preparing for the duties described in the muster list for the particular fire emergency being simulated.

1-20. Every crewmember must be given instructions that include, but are not limited to--

- The operation and use of the vessel's inflatable life rafts, the problems of hypothermia, first aid treatment for hypothermia, and other appropriate first aid procedures.
- Any special instructions necessary for use of the vessel's lifesaving appliances in severe weather and severe sea conditions, and the operation and use of fire-extinguishing appliances.

Records

1-21. When musters are held, details of abandon-ship drills, fire drills, drills with other lifesaving appliances, and onboard training must be recorded in the vessel's official logbook. Logbook entries must include at a minimal the date and time of the drill, muster, or training session, the survival craft and fire-extinguishing equipment used in the drill or drills, Identification of inoperative or malfunctioning equipment and the corrective action taken, identification of crewmembers participating in drills or training sessions, and the subject of the onboard training session.

1-22. If a full muster, drill, or training session is not held within the appointed time, an entry must be made in the logbook stating the circumstances and the details of why the event did not take place.

1-23. Once a month, if a vessel carries immersion suits or anti-exposure suits, the suits must be worn by crewmembers in at least one abandon-ship drill. If wearing the suits is impracticable due to warm weather, the crewmembers must be instructed on their donning and use.

Operational readiness

1-24. Before the vessel leaves port and at all times during the voyage, each lifesaving appliance must be in working order and ready for immediate use.

Monthly inspections

1-25. Each lifesaving appliance, including lifeboat equipment, must be inspected monthly to make sure the appliance and the equipment are complete and in good working order. A report of the inspection, including a statement as to the condition of the equipment, must be recorded in the vessel's official logbook. Each Emergency Position Indicating Radio Beacon (EPIRB) and each Search and Rescue Radio Transceiver (SART), other than an EPIRB or SART in an inflatable life raft, must be tested monthly. The EPIRB must be tested using the integrated test circuit and output indicator to determine that it is operative.

Annual inspections

1-26. Annual inspections must include the following:
- Each davit, winch, fall, and other launching appliance must be thoroughly inspected and repaired, as needed, once each year.
- Each item of survival equipment with an expiration date must be replaced during the annual inspection if the expiration date has passed, or if items in kit sets or sub-components to equipment will expire during the coming year. Date the next service / inspection as due when sub items to equipment will expire.
- Each battery clearly marked with an expiration date and used in an item of survival equipment must be replaced during the annual inspection if the expiration date has passed. If items in kits, sets or sub-components to equipment will expire during the coming year, date the next service/inspection as due when sub items to equipment will expire..
- Except for a storage battery used in a rescue boat, each battery without an expiration date that is used in an item of survival equipment must be replaced during the annual inspection.
- Rescue boat release gear must be operationally tested under a load of 1.1 times the total mass of the lifeboat when loaded with its full complement of persons and equipment whenever overhauled or at least once every 5 years.

1-27. Each vessel will have a standard operating procedure (SOP) on board that specifies Tests, Drills and Inspections (TDI). Frequencies of the TDIs are in accordance with AR 56-9, Table 2-1, Tests, Drills, and Inspections:
- ON DEMAND
  - Vessel pre-sail Checks
  - GMDSS Equipment Pre-Departure Test
  - POL Transfers to/from Vessel

- Garbage Removal
- Passenger Safety Briefing (prior to getting underway when carrying passengers)
- WEEKLY
  - Emergency Power and lighting Test
  - General Alarm Test
  - Ships Whistle Test
  - Fire Drill
  - Abandon Ship Drill
  - Person Overboard Drill
  - Underway Logbook Entries
- MONTHLY
  - Exposure suit/personal flotation device drill
  - Emergency positioning indicating radio beacon test
  - Portable fire extinguishers inspected
  - Portable dewatering pump test
  - Confined space entry meter calibrated and inspected
  - Emergency generator two hour load test
  - Search and rescue transponder test
  - Portable eyewash station and inspection
  - Ships sanitation inspection
- QUARTERLY
  - Line throwing device test
  - Breathing Apparatus inspection
  - Immersion suit inspection
  - Rescue boat test
  - Portable blower test
- SEMI-ANNUALLY
  - Emergency batteries test
  - Life ring water lights test
  - Litter slings load certification renewal
  - Battery operated flashlights and battle lanterns test
- ANNUALLY
  - Personal flotation device and attachments inspection
  - Fire main pressure test
  - Fire and foam hoses pressure test
  - Fire/smoke detection system inspection
  - Ground tackle inspection
  - Crane load test
  - Rescue boat slings test
  - Survival craft transmitter test
  - Magnetic compass deviation table renewal
  - Fixed fire extinguishing system inspections
  - Fire fighter's ensemble inspection
  - Life rings inspections
  - First aid kits inspection and component renewal
  - Commercial life raft certification and renewal
  - Commercial life raft hydrostatic release replacement and certification

- Emergency breathing devices inspection
- Confined space entry test
- Galley range extinguishing system certification renewal
- Load line certification renewal and inspection
- Military pyrotechnics serviceability inspection
- Fuel transfer hose hydrostatic test
- BI-ANNUAL
  - Remote control valves test
  - EPIRB hydrostatic release replacement and certification
  - EPIRB registration certification renewal
  - Foam tanks contents test
  - Sprinkler systems test
  - NBC sprinkler system test
- TRIENNIALLY
  - Radio frequency authorization certification renewal
- 5TH YEAR
  - Navy life raft certification renewal
  - GMDSS battery test
  - Hydrostatic test on portable fire extinguisher CO2 bottles
- 6th YEAR
  - Portable fire extinguisher remanufacture/replacement (dry chemical)
- 12th YEAR
  - Compressed gas bottle hydrostatic certification (maximum)

Passenger Safety Briefing

1-28. Passengers and special personnel must be instructed in the use of the lifejackets and the actions to take in an emergency. Whenever new passengers or special personnel embark, a safety briefing must be given immediately before sailing or immediately upon setting sail. The announcement must be made on the vessel's public address system or by other equivalent means likely to be heard by the passengers and special personnel. The briefing may be included in the muster if the muster is held immediately upon departure.

Posted Documents and Required Publications

1-29. Every Class A vessel will carry on board all Department of the Army (DA) regulations, technical manuals (TM), technical bulletins (TB), and field manuals (FM) cited in AR 56-9, Appendix A

1-30. Every unit with assigned Class B or C vessels will maintain publications cited in AR 56-9, Appendix A.

1-31. Documents will be posted as specified in Table 1-1.

Table 1-1, Document Posting Areas

Document	Posting Requirement
Fire Control and Emergency Equipment plan	Engine room Common areas Bridge
Load Line Certificate	Bridge
Lifesaving Signals	Bridge Common areas
Station Bill (complete and current)	Engine room Common areas Bridge
Pollution Placards	Engine room Common areas
Fixed Firefighting System warnings, instructions	At fire fighting equipment
Emergency Steering Instructions	On Bridge and at Emergency Steering Station.
Fixed Foam Extinguishing Systems	At activating remote station.
Officers Licenses	Bridge
Radio Frequency Authorization	On the bridge near main radio transmitter
Two GMDSS Radio Operator's Certificate/ Endorsement	Bridge
GMDSS Operators Guidance for Ships in Distress (IMO 969E)	Bridge **ONLY** Ocean Going Vessels On bridge near main radio transmitter
Wheelhouse Poster	Bridge

LOGS AND RECORD BOOKS

1-32. Log and Record books appropriate for the craft are maintained on each vessel. This includes, but is not limited to Deck and Engine Logs, Communications logs, Trash logs and Oil Record Books. Actions will be recorded as required in AR 56-9, Chapter 6. This insures accountability for vessel missions, and for regulatory requirements and provides an historical record. If scheduled Tests, Drills and inspections cannot be conducted on an operational vessel for any reason, it must be documented in the log.

FIRE CONTROL PLANS

1-33. Fire control plans provide the vessel firefighting teams with detailed information of vessel layout. This information gives the damage control (DC) team strategic options for an aggressive interior attack during a ship board fire. There are standard icons used in the development of a Fire control plan aboard watercraft. See Figure 1-1 for the Master Firefighting and Safety Equipment List currently used aboard Army Watercraft.

PILOT CARD

1-34. This card is required by IMO Resolution A.601(15) and provides the information mandated by CFR for information exchange between a vessel master and pilot. The pilot card presents in a brief form the current conditions of the ship with regard to its propulsion, steering equipment and loading conditions. The pilot card can be locally produced, and should be filled out by the ship's master prior to arrival of the pilot onboard.

WHEELHOUSE POSTER

1-35. The wheelhouse poster is required by IMO Resolution A.601(15) and 33 CFR 164.35. It provides more complete information concerning ship hull and engine characteristics than the pilot card. It contains information on stopping distances, turning diameters, and trajectories for entering turns at the maximum rudder angle in loaded and ballasted conditions. Use of the standardized wheelhouse poster is helpful in presenting the required information in a form that is readily recognizable by operating personnel and pilots unfamiliar with the vessel.

SYMBOL	DESCRIPTION	SYMBOL	DESCRIPTION
	AFFF CANS		LINE THROWING DEVICE
	AFFF TANKS		RING BUOY WITH LIGHT
	EMERGENCY FUEL SHUT OFFS		RING BUOY WITH SMOKE AND LIGHT
	EXPOSURE SUIT		RING BUOY WITH LINE
	FIRE STATION, 1.5"		DC LOCKER
	FIRE STATION, 2.5"		PORTABLE BLOWER
	FIREMANS OUTFIT		PIPE REPAIR KIT
	EMERGENCY POSITION INDICATOR RADIO BEACON (EPIRB)		ELECTRICAL REPAIR KIT
	FIRE EXTINGUISHER PORTABLE		SHORING
	FIXED FIRE FIGHTING SYSTEM PULL STATION		MAUL
	FIXED SYSTEM STORAGE AREA, HALON		MSA METER (CONFINED SPACE ENTRANCE)
	FIXED SYSTEM STORAGE AREA, CO2		P-100/P-250
	FIXED SYSTEM STORAGE AREA, FM200		SUBMERSIBLE PUMP
	GENERAL ALARM BELL		FIRST AID KIT
	FIRE DETECTION PANEL		WATER JEL (BURN DRESSING KIT)
	FIRE PUMP		FIXED FIRE FIGHTING SYSTEM SIREN
	FIRE PUMP (REMOTE START)		GENERAL ALARM CONTACT MAKER
	FIRE ALARM PULL STATION		NEIL ROBERTSON STRETCHER
	FIRE AXE		STOKES LITTER
	RESCUE BOAT		GENERAL ALARM RED STROBE LIGHT
	FOG APPLICATORS, 4 FT.		FOAM STATION
	FOG APPLICATORS, 10 FT.		FIRE DAMPERS
	FIRE/FOAM MONITORS		FIRE DOOR/CRANK
	SPRINKLER VALVES		ALARM YELLOW STROBE LIGHT (FIXED FIRE FIGHTING SYSTEM)
	REMOTE FIRE DOOR CRANK		REMOTE LUBE OIL TANK SHUT OFF
	VENT FAN SHUT OFFS		HF METER
	ESCAPE ROUTES		INTERNATIONAL SHORE CONNECTION
	OBA		PUMP HOSES
	DISTRESS SIGNAL		REMOTE ENGINE SHUT OFF
	INFLATABLE LIFERAFT		REACH ROD SHUT OFF
	LIFE PRESERVERS		VERTICAL LADDER
	LIFERAFT RADIO		SPACE PROTECTED WITH A SPRINKLER SYSTEM
	LIFERAFT TRANSPONDER		AFFF CONTROL MANIFOLD

Figure 1-1, Master Firefighting & Safety Equipment List (under revision)

STATION BILL

1-36. Army watercraft use standardized emergency signals beyond the legally required signals to communicate both interior to the vessel and external to other vessels. The following are the specific signals and instructions which are part of the vessel's Station Bill which is required to be posted in various areas throughout the vessel.

SIGNALS	
FIRE AND DAMAGE CONTROL--------	Continuous sounding of the ship's whistle and the general alarm for not less than 10 seconds.
BATTLE STATIONS------------------------	Multiple Short Rings for a period of at least 30 seconds on the General Alarm followed by the announcement "General Quarters, Man Your Battle Stations"
MAN OVERBOARD------------------------	Hail and Pass the word to the bridge, raise international code flag "Oscar" and pass the announcement "Man Overboard"
ABANDON SHIP----------------------------	More than 6 short (1 second) blasts and 1 long blast (not less than 10 seconds) on the ship's whistle and general alarm. Followed by the announcement "Abandon Ship"
SEA AND ANCHOR DETAIL--------------	The passing of the announcement "Set Sea and Anchor Detail"
NBC STATIONS------------------------------	Multiple Short Rings for a period of at least 30 seconds on the General Alarm followed by the announcement "NBC Stations"
COLLISION----------------------------------	Sounding of 5 short rings on the ship's General Alarm and whistle followed by the announcement "Prepare for Collision"

INSTRUCTIONS

GENERAL:
1. Establish Personnel Accountability.
2. Entire crew shall familiarize themselves with the location and duties of their emergency stations immediately upon reporting aboard.
3. Each crewmember shall be provided with an individual supplementary station bill card, which must show in detail the specific duties to perform.
4. Entire crew shall be instructed in the performance of their specific duties and crew on watch will remain on watch until properly relieved.
5. Emergency signals shall be supplemented with specific directions given on the public address system.

FIRE AND DAMAGE CONTROL:
1. Emergency squads will assemble at designated areas with their personal protective equipment to respond to fire or damage control.
2. Person discovering FIRE shall immediately notify the bridge by sounding the nearest alarm and fight the fire with available equipment.
3. Start fire pumps. Close all watertight doors, fire doors, ports and air vents. Stop all fans and blowers. Secure Air Conditioning Plant. Start the emergency generator.

SEA AND ANCHOR DETAIL/BATTLE STATIONS/NBC STATIONS:
1. Entire crew will report to their designated stations.
2. Each station will notify the bridge when manned and ready.
3. When Battle Stations is set, the entire crew will report to their stations with assigned weapon and NBC Gear on hand and at MOPP Level 0 unless otherwise specified in the announcement.
4. MOPP Levels will increase / decrease by direction of the Vessel Master through the announcement system.

MAN OVERBOARD:
1. Hail, and pass the word, **MAN OVERBOARD**, to the bridge. Throw life rings in water.
2. Establish personnel accountability.
3. Post lookouts. Maneuver vessel to recover person. Prepare to launch rescue boat. Hoist Oscar flag.

ABANDON SHIP:
1. All persons shall don their life preserver or exposure suit as directed.
2. Establish personnel accountability.

COLLISION:
1. Close all watertight hatches and standby for violent vessel maneuvering.
2. Prepare to perform damage control duties or abandon ship.

Figure 1-2, Station Bill Instructions

MANEUVERING CHARACTERISTICS

1-37. This information is normally available in the following three documents:

- **Pilot Card.** This card provides in brief form, the current conditions of the ship with regard to its propulsion and steering equipment and loading conditions of the ship. The pilot card should be filled out by the ship's Master prior to the arrival of the pilot. A standardized format for the pilot card will benefit all parties involved and can prevent omission of important information when briefing the pilot.

- **Wheelhouse Poster.** The wheelhouse poster provides more complete information concerning ship hull and engine characteristics than the pilot card. It also contains information on stopping distances and trajectories for entering turns at the maximum rudder angle in loaded and ballasted conditions. Use of these standardized wheelhouse posters is helpful in presenting the required information in a form that is readily recognizable by operating personnel and pilots unfamiliar with the vessel.

- **Maneuvering Booklet.** A Maneuvering Booklet may also be developed to provide detailed information about the ship's maneuvering characteristics in different conditions.

Load Line Certification

1-38. In general, most commercial U.S. vessels more than 79 feet (24 m) in length must have a valid load line certificate when venturing outside the U.S. Boundary Line, whether on a domestic or international voyage (even on *"voyages to nowhere"* that return to the same domestic port of departure). As per AR 56-9, Army vessels will maintain load line certificates. This requirement cannot be waivered by any service or component.

1-39. Load line requirements, set forth in 46 Code of Federal Regulations (CFR) 42, Subchapter E, are the basis for locating load line marks on a vessel. These marks, affixed to the vessel amidships, indicate the maximum drafts to which the vessel can be legally loaded under prescribed conditions. The distance measured vertically at the side of a vessel from the edge of the so-called "freeboard deck" to the upper edge of a particular load line mark is called "statutory freeboard" -- the "minimum statutory freeboard" measured to the uppermost load line mark applicable for a specified set of conditions taking into account considerations of 1) reserve buoyancy (buoyancy which can be supplied by the hull and watertight superstructure above the water line) and height of weather deck above this water line, 2) subdivision, and 3) hull strength. In the United States, the American Bureau of Shipping (ABS) is the load line assigning authority on behalf of the U.S. Coast Guard.

1-40. Load lines information is given in the vessel's "Load Lines Certificate." This document certifies to the correctness of the load line marks and that the vessel is in compliance with all applicable requirements. It also provides a diagram of the assigned load line marks and the freeboard deck line, locating the marks with reference to this line in terms of assigned freeboard, as well as stating any conditions, restrictions and exemptions that the vessel shall observe. The validity of these certificates is reviewed annually in load line inspections and every five years in more thorough load line surveys. Annual Load-line Certificate inspections not conducted can cause the certification to be suspended or revoked. During these inspections and surveys, ABS is particularly concerned with the following items:

- Freeing ports - Drainage must be adequate from all weather deck areas and not blocked. Particular attention is given to potential water-trapping areas such as wells formed by structure or pockets formed by cargo or equipment.

- Sill heights - Access openings in superstructure and deck houses must have 15 - inch sills. A reduction of one inch in sill height is allowed for each foot of excess freeboard with a minimum height of 6 inches.

- Vent and hatch coaming heights and fittings above the assigned freeboard deck are carefully checked.

- Watertight doors and fittings - Any penetration of watertight boundaries must be as high and as far inboard as possible. As a minimum, three dogs are required on a circular fitting and four on an oblong fitting.

- Subdivision in general - Subdivision requirements must be met as applicable for vessels being inspected / surveyed. These requirements are the same as for those passenger vessels carrying 400 or fewer passengers and include provisions for a collision bulkhead.

- A load line map showing zones and seasonal areas of the world's oceans provides the Master with information regarding the maximum draft amidships to which his vessel can be loaded during various segments of a cruise. The vessel must be loaded at the beginning of a cruise so that at no time during the cruise will, the applicable seasonal/zone mark, be submerged.

1-41. Freeboard is vitally important on smaller vessels, which are not subject to load line requirements. Consequently, these vessels should carry information on board regarding maximum drafts amidships to which they can be loaded safely.

TRAINING

1-42. The Marine Safety Inspector classes are incorporated into the Transportation Warrant Officer Advanced Course for MOS' 880A and 881A. Safety Training is provided to Army Mariners during the Warrant Officer Basic Course (WOBC), Marine Deck / Marine Engineering Warrant Officer 880A/881A A2 Certification Course (MDO/MEO A2CC), and Warrant Officer Advanced Course (WOAC) in subject areas of Army Accident Reporting, Composite Risk Management (CRM), Rules and Regulations, General Shipboard Safety, and Safety Survey Procedures.

WATERCRAFT COMPOSITE RISK MANAGEMENT

1-43. **Background.** Leaders must develop techniques that will conserve and preserve resources. Because the Army operates worldwide, missions have become increasingly demanding and so have the risks inherent in those missions. This increase in risks requires leaders to balance reasonable risks with essential mission needs. Refer to FM 5-19 *Composite Risk Management*. DA Pam 385-30 *Mishap Risk Management*, further defines application of CRM matrices to manage risk. See Figure 1-2.

1-44. **Definition.** Risk is the possibility of loss. See Figure 1-3. The loss can be death, injury, property damage, or mission failure. Composite Risk Management (CRM) identifies risks associated with a particular operation and weighs those risks against the overall mission value to be gained. The four principles of risk management are—

- Accept no unnecessary risk.
- Accept risks when benefits outweigh losses.
- Make risk decisions at the proper level (consistent with local command policy).
- Manage risk in the concept and planning stages whenever possible.

1-45. **Risk** management **process**:

- Identify hazards.
- Assess the risk of those hazards.
- Consider control options and make decisions.
- Implement controls.
- Supervise.

Table 1-2 Risk Management Matrix					
	HAZARD PROBABILITY				
	A	B	C	D	E
HAZARD SEVERITY	Frequent	Likely	Occasional	Seldom	Unlikely
I Catastrophic	Extremely High (1)	E (1)	H (2)	H (2)	M (3)
II Critical	E (1)	H (2)	H (2)	M (3)	L (4)
III Marginal	HIGH (2)	M (3)	M (3)	L (4)	L (5)
IV Negligible	MEDIUM (3)	LOW (4)	L (4)	L (5)	L (5)
NA – UNKNOWN		NH - NOT A HAZARD			

Table 1-3 Risk Management Factors		
Description	**Category**	**Outcome**
HAZARD PROBABILITY CATEGORIES		
Frequent	A	Likely to occur frequently
Likely	B	Will occur several times in the life of an item
Occasional	C	Likely to occur sometime in the life of an item
Seldom	D	Unlikely, but possible to occur in the life of an item
Unlikely	E	So unlikely, it can be assumed occurrence may not be experienced
HAZARD SEVERITY CATEGORIES		
Catastrophic	I	Death or system loss
Critical	II	Severe injury, severe occupational illness, or major system damage
Marginal	III	Minor injury, minor occupational illness, or minor system damage
Negligible	IV	Less than minor injury, occupational illness or damage

This page intentionally left blank.

Chapter 2

LIFESAVING EQUIPMENT

PERSONAL FLOTATION DEVICES

This section contains information that is common to all types of Personal Flotation Devices (PFD). It covers recommended use, required inspections, maintenance, and modifications for each type of PFD. Table 2-1 describes the types of PFD and their maximum buoyancy.

TYPE I LIFE PRESERVERS

2-1.   The Type I PFD is the primary PFD used by Army personnel aboard watercraft when mobility is not a factor. Its use by personnel aboard Army watercraft is mandatory during such procedures as abandon ship and general quarters (except aboard ships equipped with immersion suits) and general quarters. It is also mandatory for personnel on exterior weather decks and during towing operations during heavy weather. An important feature of this PFD is its ability to hold the head of an unconscious person face up (except when worn with hypothermia protective garments like anti-exposure coveralls). WARNING: Personnel who fall into the water from a great height may be initially stunned or injured. The Type I preserver is important where risk of neck injury is present. This PFD allows a person to relax completely, extending survival time and allowing the person to assume a position to protect the body from hypothermia. The main disadvantages of this PFD are its bulk (which restricts freedom of movement) and its minimum inherent buoyancy (about 22 pounds) which hampers egress from a capsized boat or swimming under water to avoid burning oil.

2-2.   Personnel should wear PFDs aboard Army watercraft when risk of an individual falling overboard exists. When assessing risk, the commanding officer or master/coxswain should consider factors such as vessel size, time required to recover a person overboard, water and air temperature, sea and weather conditions, and the degree of mobility necessary for personnel to complete a task.

NOTE: The Type I is the primary PFD for abandon ship procedures.

2-3.   The nonreversible vest style Type I PFD is designed for comfort and performance. A pass-thru slot in the rear panel allows the user to wear a safety harness beneath the life jacket and secure with a tether. Features soft, comfortable Aqua-foam flotation and provides a minimum of 22 lbs. of buoyancy.

The Type I Life Preserver can be used with a safety harness if required. The use of the harness and PFD may be of benefit to personnel working over the side from a great height or personnel working topside during heavy weather. The user will properly don a safety harness inserting the "D-ring" in the pass-thru slot in the rear panel provided in the life jacket.

Table 2-1. Types of PFDs	
Type PFD	Minimum Adult buoyancy in Pounds
I – Inflatable	33.0
I – Buoyant Foam or Kapok	22.0
II – Inflatable	33.0
II- Buoyant Foam or Kapok	15.5
III – Inflatable	22.0
III- Buoyant Foam	15.5
IV Ring Buoys	16.5
IV Boat Cushions	18.0
V Hybrid Inflatable	22.0 (Fully inflated) 7.5 (Deflated)
V Special Use Device – Inflatable	22.0 to 34.0
V Special Use Device - Buoyant Foam	15.5 to 22.0

NUMBER REQUIRED

2-4. All watercraft will carry a standard Type I life preserver for each authorized person on board. An additional number shall be provided for personnel on watch in the engine room, pilothouse, and for the bow lookout. Watercraft with living/working spaces forward, separated from messing or recreational spaces, shall stow additional life preservers for 50 percent of the total number of crew members on board in those spaces. It is the responsibility of the commands being supported, not the vessel or their command, to provide life preservers for any passengers who embark on board Army watercraft other than the Theater Support Vessel (TSV) or Landing Craft, Mechanized (LCM) 8 MOD 2 passenger carrying vessel. More compact, easily storable Type I Life preservers may be carried for passenger use, in lieu of the more bulky deluxe model.

2-5. Additionally, class "A" watercraft will carry one immersion suit for each assigned crew member on board, plus an additional immersion suit for each underway watch station (such as bridge, lookout, and engine room).

2-6. Make the following modifications before placing the PFD into service.
  ● All PFDs placed into service will have a whistle, one distress signal light, and reflective tape/material secured to the vest.
  ● Check whistle and distress signal light for proper operation. (If distress light is the chemical type, DO NOT activate until required.)
  ● Mark the vest with the vessel's name or hull number.

MAINTENANCE, INSPECTION AND TEST REQUIREMENTS

2-7. Complete the following inspections and tests as required:

Quarterly Inspection.
  ● Inspect the PFD for tears, rips, and missing webbing, tapes, and hardware.
  ● Inspect and test the whistles and distress signal lights. (DO NOT activate chemical light. Check for expiration date.)

- Inspect the reflective tape/material for cracking, peeling, and discoloration.
- Replace as necessary.

Semiannual Inspection.

- Complete all quarterly inspection steps as described above.
- Tug sharply on all straps and ties to check for rotted fabric or broken stitching.
- If any of the above does not meet standard, replace the vest.

NOTE: Before placing the PFD into service, perform and record in the logbook a semi-annual inspection as described in above paragraph.

STOWAGE

2-8. PFDs should be stored in a dry place out of direct sunlight. Heat, moisture, and light contribute to the deterioration of the PFD. Duplicate PFDs are required for persons whose normal workstation is not near their berthing area. Immersion suits are intended for "abandon ship" use. Stow them so they are readily accessible to the individuals for whom they are intended with container handles exposed, or according to manufacturer's directions. This is to prevent searching throughout the vessel to find them in an emergency. Ensure suit is dry and clean. Do not stack suits. Excessive stacking can compress suits at the bottom of the pile, eventually damaging the buoyant insulating foam. Keep all PFDs away from oil, paint, and greasy substances. *DO NOT STOW ANY PFD INSIDE A LOCKED CONTAINER.*

2-9. PFDs are equipped with reflective tape/material when they are manufactured. The material is positioned on the suit to make a person wearing the suit in the water as visible as possible under nighttime search conditions. The pattern is not necessarily the same as that used on a lifejacket or other PFD. Remove and replace unserviceable reflective tape/material by cutting two 2 X 4-inch pieces of reflective tape/material and applying it to the PFD.

Donning and Adjusting

2-10. To don and adjust the Type I, Life Preserver complete the following steps:

- Don the Type I, Life Preserver as you would any vest or shirt.
- Secure the adjustable encircling body belts and chest strap.
- Adjust the front vertical straps to provide for a snug fit.

TYPE III WORK VEST

2-11. The lightweight Type III PFD includes soft comfortable closed cell foam inside a heavy-duty shell and encircling body belt with snag resistant buckle for quick adjustment. The hinged back panel and mesh lining for ventilation gives maximum comfort in a work vest. This vest also includes a flotation collar for added protection and 62 sq. in. of reflective tape. This design allows the PFD to conform more closely to the shape of the body. Because this vest-type PFD is light in weight, the wearer can work in comparative comfort.

2-12. The Type III PFD provides less flotation than the Type I. They will not hold the head of an unconscious person face up to ensure survival. Their use may be appropriate when greater freedom of movement is needed and the risk of falling into the water from a great height does not exist. People on floats consider them acceptable for use if safety rails are in use and over the side on stages or boatswain's chairs if the person is secured by a tended safety line. These PFDs may be used aboard Army watercraft in calm weather and in calm water. Their main disadvantages are limited flotation, the tendency to ride up on the wearer, minimum buoyancy (about 16 to 18 pounds), and requires conscious effort to keep the wearer's head out of the water.

> **WARNING**
>
> The work-type PFD is buoyant enough to keep the wearer afloat in calm conditions. It has no self-righting capability and will not keep an unconscious wearer's head out of the water while awaiting rescue. This PFD is to be worn when the water temperature exceeds 60° F.

Donning and Adjusting

2-13. To don and adjust the Work Vest complete the following steps:
- Don the Work Vest as you would any vest or shirt.
- Secure the adjustable encircling body belts and chest strap.
- Ensure the reflective tape is visible.

HYPOTHERMIA PROTECTIVE CLOTHING

This section contains information about hypothermia protective clothing designed to permit personnel to function and survive in cold water. Table 2-2 describes how hypothermia affects most adults.

Table 2-2. Hypothermia Effects

Water Temperature in Degrees F	Exhaustion or Unconsciousness	Expected Time of Survival
32.5	Under 15 min	Under 15 to 45 min
32.5 to 40	15 to 30 min	30 to 90 min
40 to 50	30 to 60 min	1 to 3 hrs
50 to 60	1 to 2 hrs	1 to 6 hrs
60 to 70	2 to 7 hrs	2 to 40 hrs
70 to 80	2 to 12 hrs	3 hrs to Indefinite
Over 80	Indefinite	Indefinite

> **WARNING**
>
> The immersion suit provides the best protection from hypothermia in the water. However, it is extremely bulky and awkward to work in and is therefore limited to use for crews operating in cold weather when abandoning ship.
>
> The anti-exposure coverall provides good durability and out-of-water protection from the elements. It provides limited protection from hypothermia to crewmembers in the water.

APPLICATION

2-14. Commanders and vessel masters will ensure compliance with the guidelines described below:

● Watercraft crewmembers shall wear hypothermia protective clothing if the **water temperature** is below 15.6° Celsius (60 degrees Fahrenheit).

● The commander or vessel master may waive the requirement for wearing an anti-exposure coverall if the degree of risk to hypothermia is small (such as in non-hazardous daylight rescue operations in calm water).

● A PFD should NOT be worn over an anti-exposure coverall or survival suit because the device is inherently buoyant. Although a PFD will improve chances for survival during prolonged periods because it provides improved flotation, the additional buoyancy creates problems for the wearer attempting to leave capsized watercraft.

ANTI-EXPOSURE COVERALLS

2-15. Personnel operating in a cold, wet environment wear the anti-exposure coverall when they need protection from hypothermia (see Figure2-1) when operating in an area where the water temperature is less than 59°F. The anti-exposure coverall (often called a "deck suit" or "work suit") affords adequate protection from exposure to cold water, wind, and spray. It provides flotation similar to that provided by the work vest.

2-16. The main advantage of the Type III PFD is its wearability, ease of donning, simple construction, and neat appearance. The disadvantages are limited flotation (no righting moment) and minimum buoyancy (about 16 pounds). Wear this PFD only when you require greater freedom of movement and the mission and environment are less hostile.

2-17. Anti-Exposure Coverall, Type III PFDs are not universally sized. Two or three different sizes are required to fit adults properly.

2-18. Type III PFDs may be used as a substitute for the work-type preserver.

NOTE: The anti-exposure coverall is primarily used by watercraft crewmembers where they may be exposed to intermittent spray.

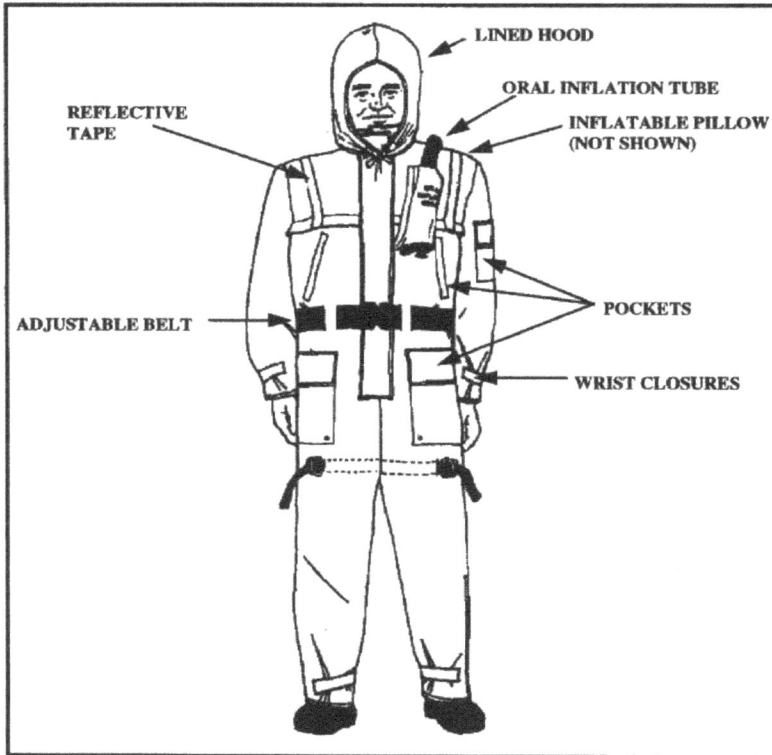

Figure 2-1. Anti-exposure Coverall

2-19. The anti-exposure coverall is made of orange urethane-coated nylon exterior fabric with a closed-cell foam interlining to provide thermal protection. It provides at least 17 ½ pounds of buoyancy. The coverall allows full freedom of movement. The suit features an attached, orally inflated pillow to support the wearer's head in the water. It also has an attached hood for extra thermal protection and reflective tape/material on the hood and shoulders for better visibility at night. For added protection, personnel should carry waterproof gloves for use with the anti-exposure coverall. The coverall is manufactured in five sizes ranging from small to extra-large.

2-20. Don anti-exposure coveralls in the same fashion as standard coveralls.

MAINTENANCE, INSPECTION AND TEST REQUIREMENTS

Cleaning

2-21. To clean the anti-exposure coverall, complete the following steps:
- When coveralls have been submerged or exposed to salt water or salt spray, wash them in a shower with a mild soap.
- Units may machine wash excessively soiled anti-exposure coveralls. Use a gentle cycle and mild soap. The water temperature should not exceed 105 degrees F.

> **CAUTION**
>
> Do NOT attempt to dry anti-exposure coveralls in a clothes dryer. Do NOT wring out anti-exposure coverall. To dry coveralls, hang on a wooden hanger in a cool, dry, well-ventilated area. Do not dry in direct sunlight.

> **CAUTION**
>
> Do not use thinners, solvents, or similar agents for cleaning coveralls that have been exposed to paint, paint removers, acids, solvents, gasoline, or any substance containing acetones.

Inspection and Maintenance

2-22. Units shall inspect the anti-exposure coverall quarterly. To inspect the coverall, proceed as follows:

- Lubricate the zipper
- Lay out suit and check for obvious damage.
- Work entry zipper up and down to check for ease of operation. Rubbing a bar of soap or paraffin (NO oil or grease) over edges of zipper will ease operation.
- Check buoyancy chamber and inflation tube for obvious damage.
- Inflate buoyancy chamber and check for leaks.
- Deflate chamber and stow in chamber casing.

Repairs

2-23. Units should make only minor sewing repairs to anti-exposure coveralls. Obtain commercial assistance for repairs beyond the capabilities of the unit.

IMMERSION SUIT

2-24. The immersion suit (also referred to as an "exposure suit") is worn by crews when abandoning ship. The suit affords protection from exposure to cold water, wind, and spray. The foam fabric is a durable and elastic material with high flotation characteristics providing approximately 35 pounds of buoyancy.

2-25. The approved immersion suit (Figure 2-2) is a one-piece, international orange garment constructed from 3/16-inch nylon-lined neoprene or polyvinyl chloride (PVC) foam.

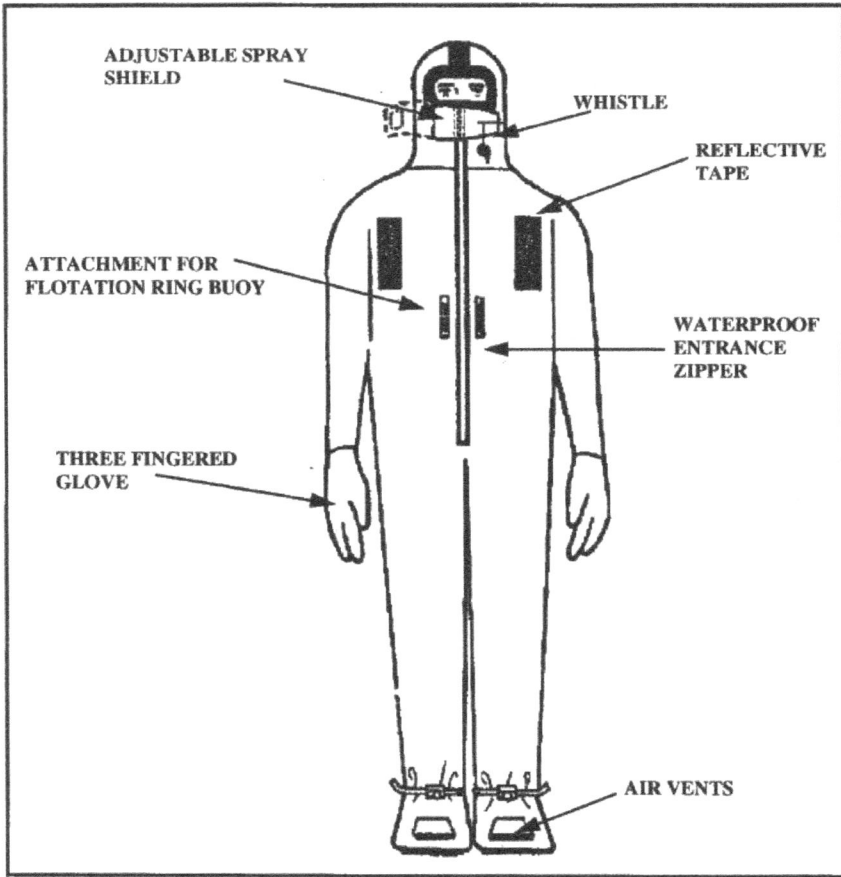

Figure 2-2. (Immersion) suit (typical)

NOTE: The buoyancy provided to the lower torso will cause the wearer to float horizontally either face up or face down in rough seas. Additional flotation, such as the inflatable collar provided with the suit, must be used to assure face up flotation. The Adult Universal survival suit is designed so that one size will fit most persons (weighing between 110 and 330 pounds). Other sizes are available. The thermal qualities of the fabric/foam laminate will keep survivors warm whether they are wet or dry.

2-26. Masters and coxswains of watercraft that have immersion suits will ensure that every other abandon ship drill conducted uses the immersion suits in lieu of the Life vest.

2-27. Attach a personal distress signal light to the left breast pocket.

2-28. To don the immersion suit, proceed as follows:

- Remove suit from stowage bag with a sharp jerk of the carrying bag.
- Don suit in the same fashion as donning coveralls.
- Don the hood before you zip up the suit.
- Close the zipper completely. To avoid problems zipping up the suit, arch your back to remove wrinkles in the fabric.
- Close the spray shield and inflate the collar for additional flotation.

CAUTION

When donning/wearing the immersion suit, extreme caution shall be taken to avoid sharp, protruding objects that may snag or tear the suit.

MAINTENANCE, INSPECTION AND TEST REQUIREMENTS

Cleaning.

2-29. To clean the immersion suit, complete the following steps:

- When suits have been submerged or exposed to salt water spray, suits shall be washed under a shower with a mild soap.
- Do NOT wring out immersion suits.
- To dry the suit, hang it on a wooden hanger in a cool, dry, well-ventilated area.
- Do NOT dry in direct sunlight.

CAUTION

In NO situation shall thinners, solvents, or any similar agents be used to clean suits that have been exposed to paint, paint removers, acids, solvents, gasoline, or any substance containing acetones

---

Inspection and Maintenance

2-30. The immersion suit shall be inspected before being placed into service and quarterly thereafter. To inspect the suit, proceed as follows:

- Stowage bag - Check condition of snaps on bag for ease of operation.
- Suit - Lay out on a flat, clean surface and check for obvious damage.
- Zipper - Work zipper up and down to check for ease of operation. If zipper is excessively rough, wipe with a soft, clean, lint-free cloth and lubricate with the wax lubricant found in suit breast pocket.

NOTE: The teeth that actually secure the waterproof zipper are the small teeth on the inside of the zipper. A little corrosion on these teeth can block the slider or damage the teeth so the zipper does not operate. If a closed zipper can be separated when probed with a (dull) knife, the zipper needs to be replaced.

- Inflatable collar - Check collar for obvious damage.

NOTE: Periodically inflate and allow it to stand overnight. If the collar does not stay firmly inflated overnight, it should be repaired or replaced. Inspect lock screw on inflatable collar inflation tube to ensure that it is in the unlocked position.

Packing

2-31. To repack the survival suit after inspection, follow the manufacturers instructions or proceed as follows:

- Lay out suit on a flat clean surface with front up and arms out.
- Make sure entry zipper is in the open position.
- Roll or fold suit, feet first, up to chin, making sure not to wrinkle water valves.
- Fold arms horizontally across roll.
- Place suit in bag and close snaps.
- Stow bag with handle exposed.

Repairs

2-32. The following repairs are authorized on the immersion suit. Commercial assistance should be obtained for repairs beyond the capability of the unit.

> **CAUTION**
>
> Repairs should be made only with neoprene (contact) cement. Other cements may contain solvents that would weaken the material.

- **Separated seams, rips and tears.** To repair, complete the following steps:
  - Trim jagged edges with scissors until new rubber shows.
  - Remove old cement.
  - Thoroughly dry material.
  - Apply neoprene cement IAW manufactures instructions:
  - Apply four coats of cement along entire surface of material to be repaired.
  - Allow each coat to dry between each application.
  - When the last coat becomes tacky, align edges. Apply firm and even pressure when pressing edges together. Hold edges together for 3 or 4 minutes.
  - Allow at least 1 hour for cement to set before using repaired item.
- **Holes.** To repair holes, complete the following steps:
  - When entire areas are missing, trim edges of area to convenient configuration.
  - Cut a replacement piece conforming to size and shape of prepared area.
  - Proceed as in steps shown in repairing separated seams above.
- **Corroded zippers.** To clean zippers, complete the following steps:
  - Scrub with toothbrush, using fresh water.
  - Rub a bar of soap or paraffin wax (NO oil or grease) over edges of zipper to act as a lubricant and retard corrosion.

PERSONNEL MARKER LIGHT

2-33. The Personnel Marker Light (PML) is a chemical light developed for use on PFD and hypothermia protective garments (immersion suit). It is used to attract the attention of search and rescue aircraft, ships, or ground parties. Once activated it provides light for approximately 8 hours. The PML is equipped with a pin-type clip and should be attached to the PFD front left shoulder.

NOTE: The PML is the only chemical light authorized for use and is intended to replace one-cell flashlights as they go out of service.

OPERATING INSTRUCTIONS

2-34. To activate the PML, firmly squeeze lever against light tube, breaking sealing band and ampoules inside light. Slide protective sleeve from PML.

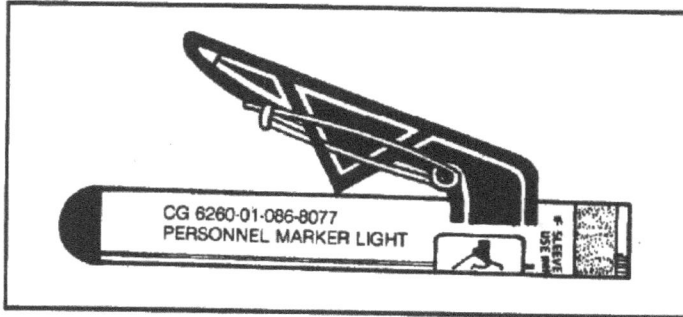

Figure 2-3. Personnel marker light

INSPECTION

2-35. Personnel Marker Lights (PML) that have been placed into service shall be inspected during normal inspection cycles of equipment, such as PFD and hypothermia protective garment, to which it is attached. Do the following to inspect the PML:

- Inspect sealing band on protective sleeve for security.
- Check expiration date stamped on sealing band. Replace PML prior to manufacturer's expiration date.
- Examine safety-pin-type clip for deformity. If pen is deformed, replace the PML.
- Ensure safety pin is in the closed position.
- Discard expended or out of date PML IAW unit HAZMAT / Environmental SOP

NOTE: Protective sleeve must be kept in place to protect against accidentally breaking the glass ampoule and to protect against deterioration of chemical by ultraviolet light.

SIGNAL WHISTLE

2-36. The signal whistle emits an audible signal to rescue ships or personnel. One whistle shall be attached to each PFD and to each hypothermia protective garment.

NOTE: Life jacket, work vest, anti-exposure suit, & survival suits include whistle attached to zipper.

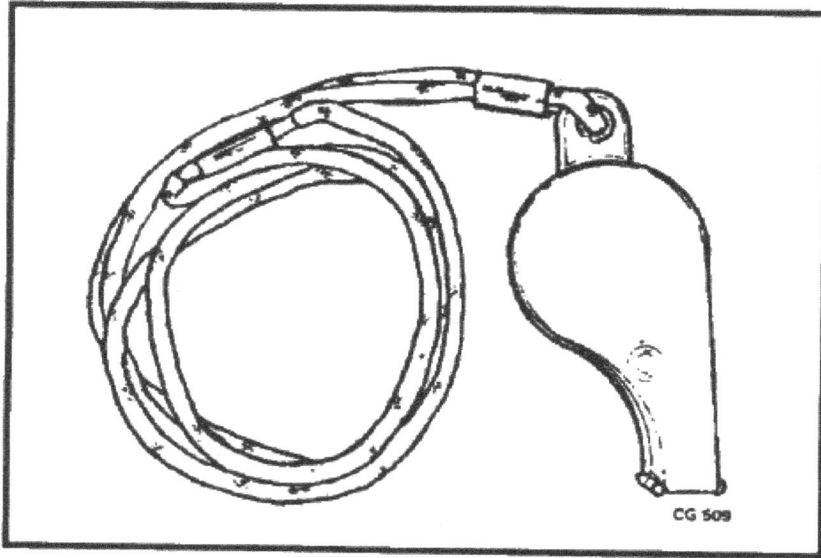

CG 509

Figure 2-4. Signal whistle

CONFIGURATION

2-37. The signal whistle (figure 2-4) is made of plastic with a lanyard attached for easy access and to prevent loss.

APPLICATION

2-38. The signal whistle is used to attract attention of rescue ships or personnel in foggy weather or at night. Whistle range is 1,000 yards.

INSPECTION

2-39. Signal whistles shall be inspected during the normal inspection cycle of the equipment (such as the PFD or hypothermia protective garment) to which it is attached. Do the following to inspect the signal whistle:

- Ensure side discs of whistle are neither loose nor missing. Check whistle for cracks and damaged ball. Replace damaged or defective whistles.
- Blow whistle normally (regular exhalation), then with forced exhalation. Replace the whistle if it fails to emit a highly audible sound.

NOTE: The signal whistle must be tethered to the PFD or hypothermia protective garment. Nylon cord shall be seared at both ends to prevent fraying. Attach the whistle to a grommet installed on the garment using a 36-inch length of Type I nylon cord. A bowline knot should be used to secure the ends of the cord to the whistle and the grommet.

RING BUOY

2-40. The standard ring buoy (figure 2-5) is a 30-inch diameter, inherently buoyant, plastic buoy. Ring buoys are intended to be used when a person falls overboard. A ring buoy with a floating distress marker attached shall be readily available on each side of the watercraft. Other life ring buoys must have a retrieving line of 90-feet, buoyant, non-kinking, 5/16 –inch diameter, 1124 lbs breaking strength and be of a dark color if synthetic, or of a type certified to be resistant to deterioration from ultraviolet light. All ring buoys will have reflective tape/material attached.

2-41. Ring buoys are deployed as follows:
- When "Man Overboard" occurs, a ring buoy with distress marker is immediately thrown towards the individual in the water. On ocean-going vessels, the ring buoy with distress marker and signal marker (smoke/flare) is deployed from the wheelhouse.
- The buoys have a two-fold purpose: 1) Mark the spot where the individual entered the water so that the area will be visible to the turning vessel, and 2) to give the individual some additional buoyancy to hold on to.

2-42. When the vessel is close enough to the individual, the ring buoy with the retrieving line is used. The bitter end of the retrieving line is either tied to the vessel or held by the soldier-mariner throwing the ring buoy to the individual in the water. After the individual gains control of this ring buoy, he is drawn to the vessel's side and hoisted on board the vessel. Reflective tape or material (figure 2-5) assists in locating the ring buoy at night.

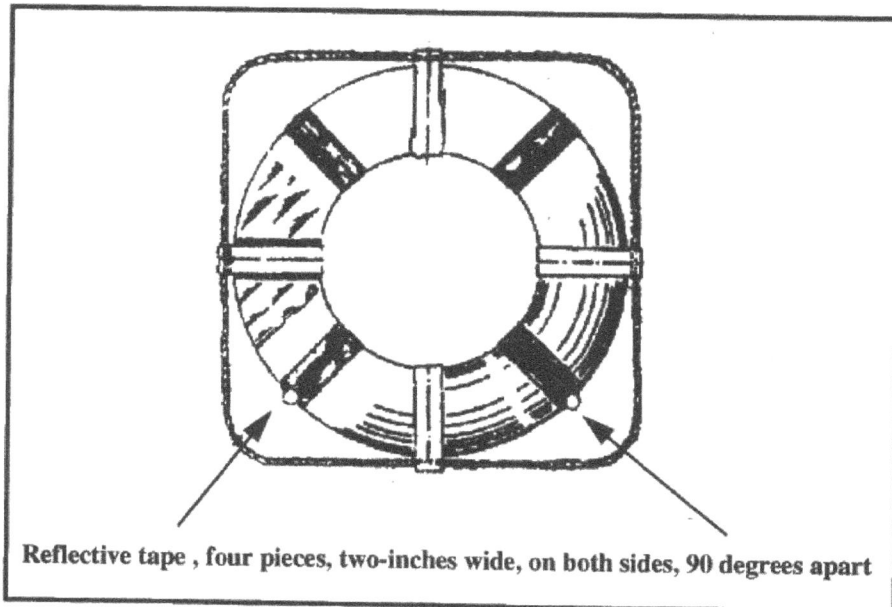

Reflective tape , four pieces, two-inches wide, on both sides, 90 degrees apart

Figure 2-5. Ring buoy with reflective tape / material

NUMBER REQUIRED

2-43. The minimum number requirements for 30-inch ring buoys and the minimum number which shall have distress markers, smoke signals, or retrieving lines, shall be in accordance with Table 2-3, page 2-15.

NOTE: Marine smoke/flame location marker will be located on the port and starboard bridge wing in the vicinity of the ring buoy with distress marker light.

Table 2-3. Required ring buoys and attachments

Type of watercraft	Minimum number of buoys required	Minimum number of buoys required with 90 foot line attached	Minimum number of buoys required with distress marker light attached	Minimum number of buoys required with marine smoke / flame location marker attached
LSV	8	2	6	2
LCU-2000	8	2	6	2
LT 800 series	6	1	5	2
LT (100 foot)	5	1	4	2
ST 900 series	5	1	4	0
LCM 8 Mod 1	3	1	2	0
LCM 8 Mod 2	3	1	2	0
BD	4	2	2	0
BG	2	1	1	0
MWT/SLWT/CF	2	1	1	0

INSPECTION

2-44. Inspect ring buoys every six (6) months as follows:
- Inspect ring buoy for general condition of inherently buoyant material, such as holes, cracks, rips, and so forth.
- Inspect condition of lifeline, replace as required.
- Inspect retrieving line for condition and security. The retrieving line shall be securely attached to the ring buoy lifeline with an eye splice. Retrieving lines shall be stowed in a loose coil and lashed in place on the ring buoy with an easily breakable cotton thread.
- Inspect the reflective tape/material for tears and missing pieces.
- Ensure that the ring buoy is orange in color (international) and stenciled with the vessel's name or number and the legend "US ARMY".
- If a ring buoy is found unserviceable, remove it from the watercraft immediately.

MODIFICATIONS

2-45. Some modifications are required for ring buoys before use:
- To aid in locating a ring buoy at night, there shall be four pieces of reflective tape / material, two inches wide, wrapped around each ring buoy at 90° apart as shown in Figure 2-5.
- Attach 90 feet of retrieving line, using an eye splice, to each ring buoy without a distress marker or smoke signal.

NOTE: Marine smoke/flame location marker will accompany the ring buoy with distress marker light located on the port and starboard bridge wing but will NOT be attached to either.

- Attach a floating distress marker to each ring buoy positioned for man overboard emergency. It is attached by the halyard snap and 2 feet of l/4-inch diameter, polyethylene line.

CAUTION

To reduce possible injuries to individuals in the water due to high heat and bright light, marker should be thrown into the water immediately after ring buoy with battery operated distress marker. Distress marker will help individual locate ring buoy while Marker, Location, Marine, Smoke/Flame (MK 58 Mod 1), will help vessel locate individual.

FLOATING DISTRESS MARKER

2-46. The floating distress marker (figure 2-6) is a watertight, vapor proof, battery-operated flashing light normally used during darkness to mark the location of objects in the water. It is attached to ring buoys used at man overboard and lookout stations. The capacitor discharge xenon flashtube emits flashes of light at a rate of 60 +/- 10 flashes per minute for a minimum duration of 15 hours. The mounting bracket is designed to hold the light in an inverted position yet release it when a pull of 20 to 40 pounds is applied. External parts of the light, other than the lens, are international orange. The floating distress marker will be modified by the crew by adding reflective tape just below the light housing when the light is in the "on" position (upright) and marking the outer casing with the vessel's hull number and "US Army".

INSPECTION AND MAINTENANCE

2-47. The floating electric marker light shall be inspected before placing the light into service and every 6 months thereafter. Maintenance of the floating electric marker light is limited to inspection, replacement of batteries, and cleaning. To inspect the light, complete the following steps:

NOTE: Weak batteries, internal corrosion, and lack of water tightness are major reasons for failure of floating lights.

- Remove light from bracket, ensuring its easy removal.
- Remove battery from light.
- Inspect lens and case for interior condensation and cracks.
- Inspect condition and security of lanyard and halyard snap hook. Replace as necessary.
- Inspect battery compartment for corrosion or signs of battery leakage. - Clean and dry all contacts.
- Clean exterior of light using a mild soap and water solution and a soft cloth.
- Thoroughly dry exterior of light.
- Replace any cracked or broken gaskets.
- Install a new battery.
- Test the light by inverting it (lens up). The light should come on and flash at a rate of 60 +/- 10 flashes per minute.
- Test internal switch by turning light upside down (lens down); light should extinguish. If light does not flash or extinguish, replace light.

> **CAUTION**
>
> Do not keep the lamp in a lighted position more than necessary since operating life will be reduced.

- Mark date of inspection and replacement of battery on outside of light using stencil, marking pen, or plastic tape.
- Ensure date of inspection is documented under Test, Drill and Inspections (TDI's) in the vessel's Official Deck Log Book.
- Replace light in bracket with lens down.

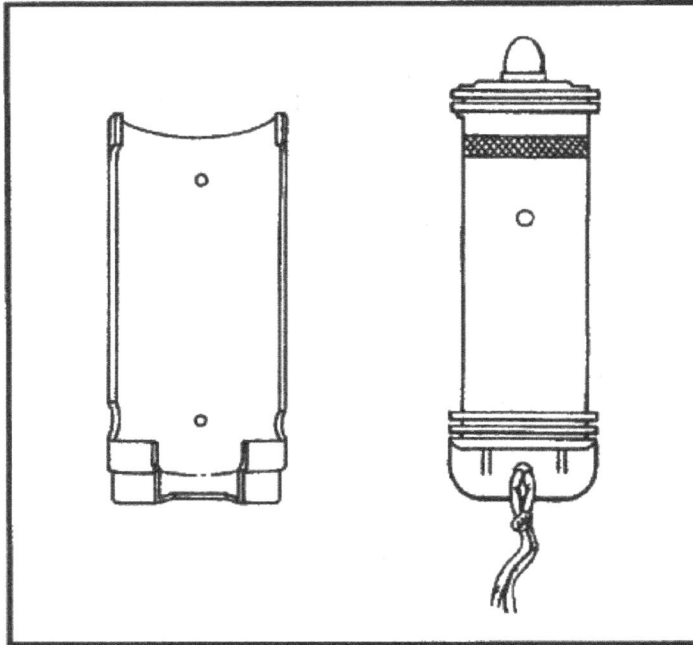

Figure 2-6. Floating distress marker

MODIFICATIONS

2-48. The following describes modifications to the floating electric marker light:

- Attach a 2-foot length of yellow, polyethylene/polypropylene, (1/4-inch diameter, 3 strand, 1000 pound breaking strength) line with an eye splice.
- Attach a halyard snap to the opposite end of the line using an eye splice. The halyard snap is used to attach the light to the ring buoy or other objects. To aid in locating a nonfunctional light, reflective tape (NSN 9390-01-078-8660) shall be applied to the upper body of the light.
- Attach reflective tape to the upper body of the light, to aid in locating a nonfunctional light.
- Ensure tape totally encircles the body of the light.

NOTE: Configuration of distress marker lights for this NSN may vary. Insure the correct battery is used by checking the owner's manual for correct requisitioning information; otherwise the light may not work or float upright when thrown into the water.

DISTRESS SIGNALS

2-49. This section contains information about military pyrotechnic distress signal devices authorized for US Army watercraft. Headquarters, Department of the Army policy states that commercial munitions, including pyrotechnics, are NOT authorized on Army equipment nor utilized by Army personnel. The only exception is in the event that a vessel is deployed forward of its home base; SOLAS standard pyrotechnics and safety items can be procured and utilized when items within the DoD procurement system cannot be obtained. The military-issue devices have no expiration date. This chapter describes the following pyrotechnic devices:

- Signal Illumination, Ground, Red Star, Parachute, M126A
- Marker, Location, Marine, Mk 58 Mod 1
- Signal, Smoke and Illumination, Marine, MK124 Mod 0

PRECAUTIONS

2-50. Precautions should be taken when using, handling, and storing pyrotechnic devices. The following are warnings, precautions, and procedures for these devices:

> **WARNING**
>
> **Personnel handling pyrotechnic signal flares shall comply with all existing safety requirements and precautions. Pyrotechnics are hazardous due to the nature of their explosive, flammable, or toxic filler.**

- DO NOT remove the signal device from its hermetically sealed container until immediately before use.
- Read and follow the firing instructions on the signal body.
- Handle pyrotechnic flares with the same care as high explosives.
- Protect flares and signals from moisture.
- Remove and replace flares when there is evidence of moisture.
- Disassembly of flares is strictly prohibited.
- DO NOT use flares when they are rusted, dented, or deformed. (They must be segregated for disposal.)
- Avoid any rough handling, throwing, or dropping of pyrotechnics.
- DO NOT look into the firing end of any signaling device.
- Remove flares and signals from watercraft placed in storage.

DISPOSAL

2-51. All unserviceable pyrotechnics should be turned in to the local Ammunition Supply Point (ASP) for proper disposal.

MISFIRES

2-52. Misfired signals must NOT be approached until at least 30 minutes have elapsed after firing was attempted. All misfires and malfunctions involving these signals will be reported through the appropriate munitions supply channels.

AUTHORIZED PYROTECHNICS ALLOCATIONS

2-53. US Army watercraft are authorized certain pyrotechnics. These pyrotechnics are described in Table 2-4.

Table 2-4. US Army watercraft authorized pyrotechnics

Type Of Watercraft	Illumination Red Star Parachute	Marker Location Marine	Smoke and Illumination
LSV	12	2	6
LCU-2000	12	2	6
LT-800	12	2	6
LT-100	12	2	0
ST-900	0	0	12
LCM-8 MOD 1	0	0	12
LCM-8 MOD 2	0	0	12
SLWT	0	0	12
BD	0	0	12
RRDF	0	0	12
CF	0	0	12

Signal, Illumination, Ground, Red Star, Parachute, M126A.

2-54. This rocket-propelled, fin-stabilized device is a hand launched distress signal for watercraft operating in ocean or coastal waters. The signal is shown in Figure 2-7.

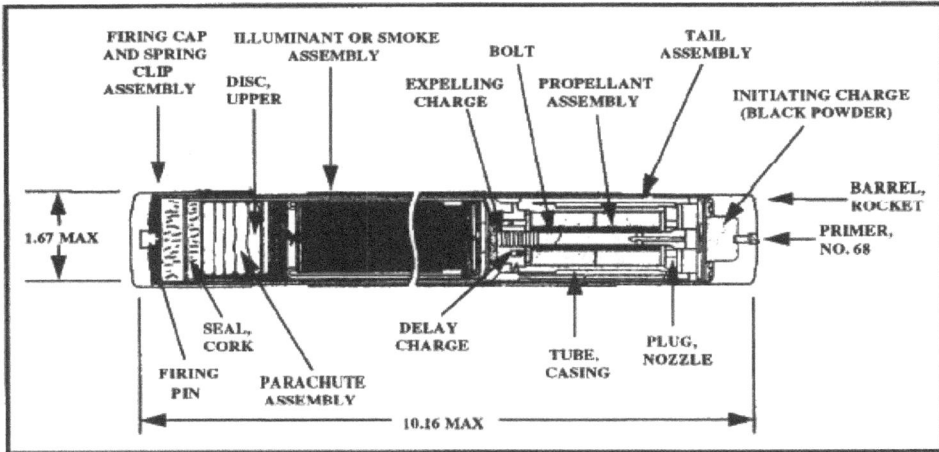

Figure 2-7. Signal, illumination red star parachute, M126A1 (Interior)

2-55. *Intended use*. When fired vertically, the signal (figure 2-8) projects to an altitude of 650 to 800 feet. It also produces a parachute-suspended red star that burns for approximately 50 seconds while descending at the rate of 8 feet per second.

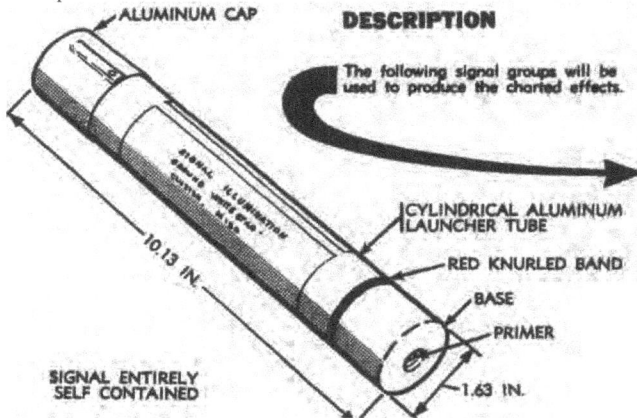

Figure 2-8. Signal, illumination red star parachute M126A1 (Exterior)

2-56. Firing Instructions. Perform the following to fire the illumination signal:

- Remove the signal in accordance with instructions printed on the container.
- Hold the signal in the left hand (red knurled band up) with thumb and forefingers in alignment with the red band (figure 2-9).

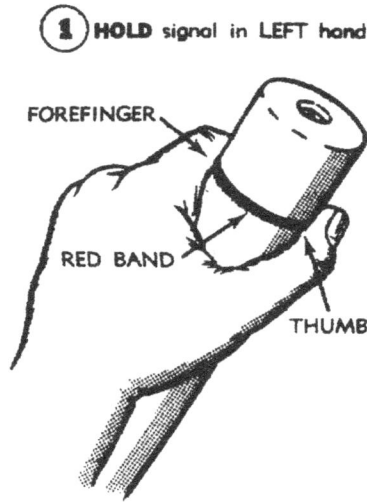

Figure 2-9. M126A1 Firing Instructions (Step 1)

- Remove the firing cap from the lower end of the signal
- Point the ejection end of the signal (the end opposite the red knurled band) away from the body and away from personnel, equipment, and materials. - - - SLOWLY push the firing cap onto the primer (red band) end until the cap is aligned with the lower edge of the knurled band (see figure 2-10).

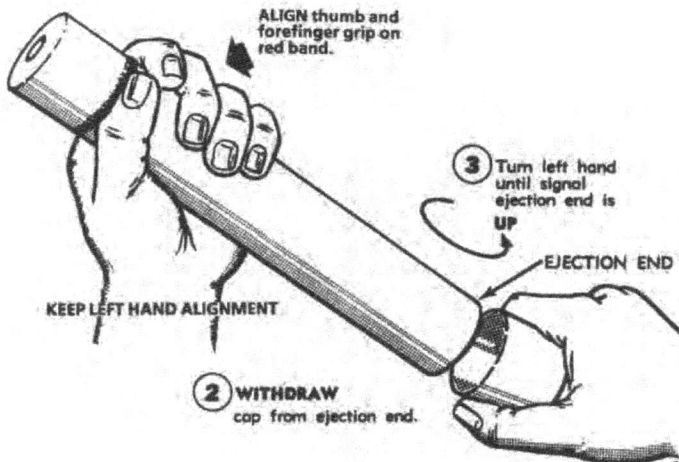

Figure 2-10. M126A1 Firing Instructions (Steps 2 and 3)

- DO NOT permit the cap to go beyond the lower edge of the band.
- Hold the signal FIRMLY at arm's length with the left hand in a vertical position (90 degrees) with the firing cap downward (figure 2-11).

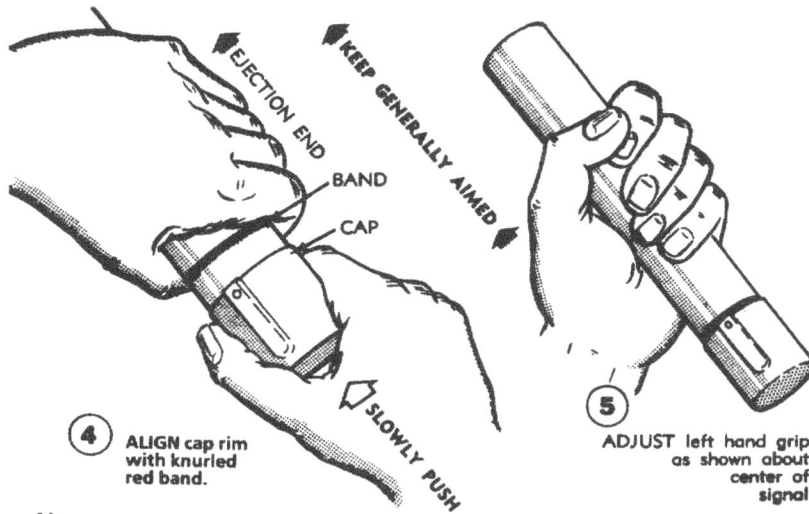

Figure 2-11. M126A1 Firing Instructions (Steps 4 and 5)

- Strike the firing cap bottom sharply with the palm of the right hand, keeping the left arm rigid (figure 2-12). Follow instructions for Misfires.

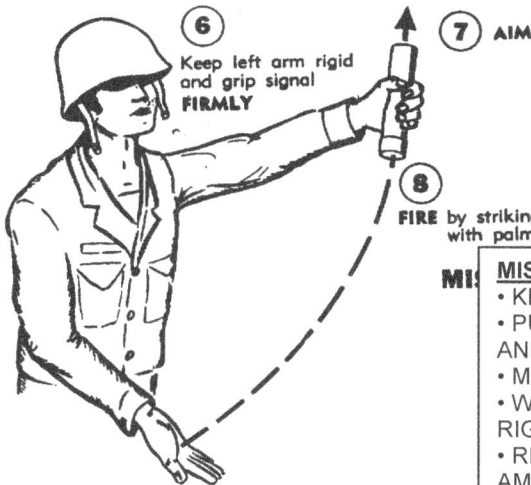

MISFIRE INSTRUCTIONS
- KEEP SIGNAL AIMED.
- PULL CAP BACK TO RED KNURLED BAND AND ROTATE 90°.
- MAKE TWO MORE ATTEMPTS TO FIRE.
- WAIT 30 SECONDS WITH ARM REMAINING RIGID AND SIGNAL AIMED OVERHEAD.
- RETURN CAP TO EJECTION END. TURN IN TO AMMUNITION SUPPLY POINT.

Figure 2-12. M126A1 Firing Instructions (Steps 6 through 8)

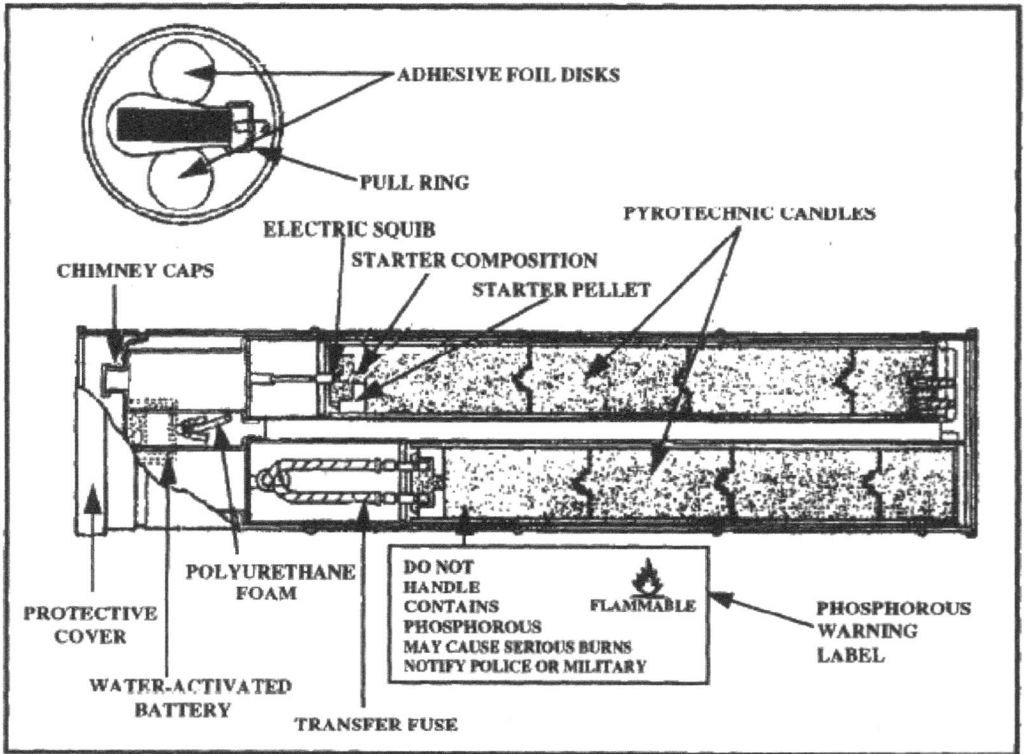

Figure 2-13. Marker Location, Marine; Mk 58 Mod 1

Marker, Location, Marine, Mk 58 Mod 1

2-57. This marker consists of a cylindrical tin can (21.78 inches long and 5.03 inches in diameter). The ignition end of the marker has three holes, two for smoke and flame emission and one for entry of water to activate the signal. Adhesive foil discs hermetically seal the two emission holes and a reinforced adhesive foil strip with a rectangular pull hermetically seals the hole for water activation. The adhesive foil seals are protected during handling and shipping by a replaceable polyethylene protective cover. A description of the marker interior parts is in Figure 2-13.

Stowage

2-58. The Marker, Location, Marine, MK 58 Mod 1. The marker will be stowed on bridge port and starboard wings in PVC container. This PVC container can be constructed by the crew. Materials needed to construct container are listed below:

- 1ea 17" piece of 6" ID PVC, schedule 40
- 2ea 6" high pressure PVC end caps
- 4ea #8 galvanized sheet metal screws
- 1ea 14" long piece of 1/16" plastic coated cable
- 2ea solder less wire connectors (for 1/16" cable)
- 1ea 4" galvanized screen door handle

2-59. The steps for fabricating the container are listed below:

- Cut 6" ID PVC to 17" in length.
- Drill 3/8" drain hole in the bottom/center of one end cap.
- Glue end cap with the 3/8" drain hole to the bottom of the 17" length of PVC.
- Fit the second end cap to the top of the 17" length of PVC (file or sand this end so the cap can be easily removed).
- Fasten the 4" screen door handle to the top end cap using 2ea #8 screws.
- Connect the solder-less wire connectors to each end of the plastic coated cable.
- Fasten the cable to the top end cap and to the side of the 17" length of PVC using 2ea #8 screws (this is to prevent losing the cap)
- Paint container gray to match the bridge exterior, stencil "SMOKE MARKER MK58 MOD 1" on the container (white letters, 1 1/2" high).
- Use non-rusting material to mount the holder, vertically with the drain hole down, on the inside of the bridge railing next to the life ring with light (ensure that the cap can be easily removed)
- Place the marine location marker in the container with the polyethylene protective cover pointing up.
- Install the top end cap.

NOTE: Condensation will build up around the base of marker and the container. Vessel crews should place a non-corrosive, non absorbent object in the base of the container to keep the marker from resting at its base on the container. Ensure that the drain hole is not blocked.

Intended use

2-60. This marker is designed for day or night use. It can be used for man overboard and any other condition calling for long-burning, smoke and flame reference point marker on the ocean surface. It produces a yellow flame and white smoke for a minimum of 40 minutes and a maximum of 60 minutes. It is visible for at least 3 miles under normal conditions.

Operation

2-61. Perform the following steps to use the location marker.

- Remove the polyethylene protective cover.
- Remove the pull ring reinforced adhesive foil strip.
- Throw the signal overboard with life ring. The signal will activate within 25 seconds of impacting the water.

SIGNAL, SMOKE AND ILLUMINATION, MARINE, MK 124 MOD 0

2-62. This signal is made of metal and cylinder-filled with illuminant composition in one end and smoke in the other. Each end is fitted with a plastic cap. The cap on the flare end has molded protrusions or beads on the face for night identification. The smoke (day signal) end cap is smooth. A label around the signal body further identifies each end and provides precise instructions for use. A description of this smoke and illumination signal is shown in Figure 2-14.

PROTECTIVE CAP   PRIMER   QUICKMATCH   FLARE CANDLE   PRIMER   IGNITER

IGNITER   FIRECRACKER FUSE   SMOKE CANDLE   PROTECTIVE CAP

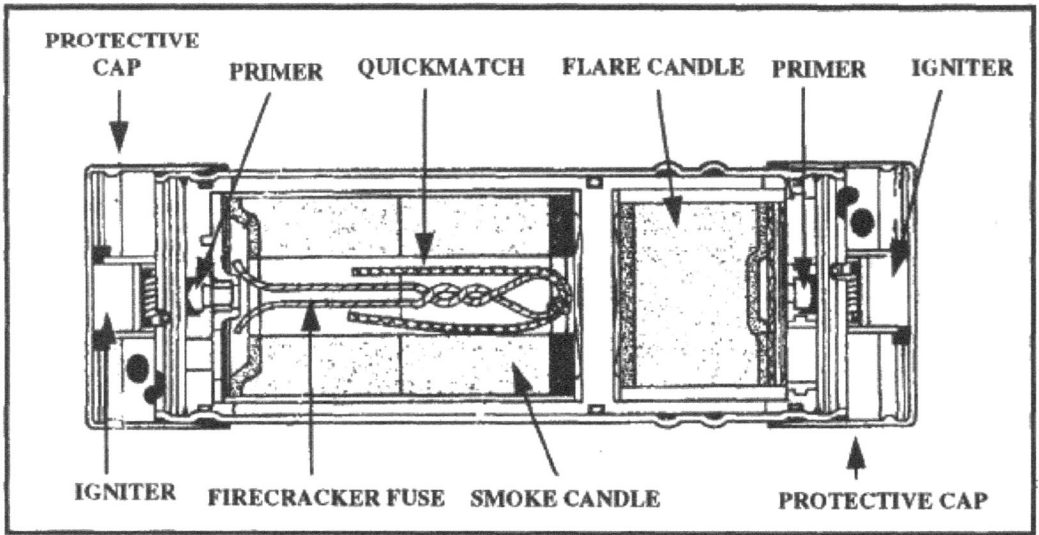

Figure 2-14. Marine Smoke & Illumination Signal, Mk 124 Mod 0

Intended use

2-63. This signal can be used for day or night signaling. The signal is a one-handed operable device intended for rescue use. Its small size permits it to be carried on all Class B Army watercraft as a primary system and on selected Class A Army watercraft as part of the vessel's Rescue Boat required equipment.

Firing instructions

2-64. After choosing the type of display desired, smoke for day or flare for night, operate the signal as follows:

WARNING

Prior to pulling the lever downward, position all fingers below top of signal.

- Remove the protective cap from the end to be ignited.
- Slide the lever horizontally to the fully extended position.
- Pull the lever downward until firing pin is released.
- If the smoke end flames, briefly immerse in water or hold against solid object.
- During and after ignition, hold signal firmly with arm fully extended overhead at an angle of 45 degrees from the body.
- DO NOT direct either end of the signal toward user or other personnel.
- After using the signal, douse the used end in water to cool. If used on land, place the signal on a noncombustible surface to cool.
- Save the signal for use of other end in case it is needed.

> # WARNING
>
> **Under no circumstances shall both ends of this signal be ignited at the same time.**

ANNUAL VISUAL INSPECTION

2-65. Pyrotechnics shall be inspected annually for current expiration date (if applicable), corrosion, dents, swelling or punctures, missing safety pins and caps, and the presence of chemical odors. Turn in defective or damaged pyrotechnics to the nearest Army supply facility.

TRAINING

2-66. No specific devices are designated for training use. To promote the safe and effective use of pyrotechnics, all units shall provide annual periods of instruction to develop and maintain proficiency and confidence in the military distress signals. The following training methods are recommended:

- Training shall cover the particular pyrotechnic item used aboard that watercraft, its manner of use, and safety precautions.
- A demonstration allows first hand observation of the device.
- At least one of each device will be expended on board. No additional expenditure is deemed necessary.
- Training will be combined with several units/watercraft to reduce the number of expended devices.
- Pyrotechnics used for training will be fired from the oldest lots on hand.
- To prevent a false sighting report, units that conduct training will notify the appropriate Harbormaster and US Coast Guard office well in advance, noting the time, place, and pyrotechnic devices to be used.

INFLATABLE LIFE RAFTS

2-67. This section contains information about inflatable life rafts used on US Army watercraft as well as life raft recertification requirements, description, location, and stowage. These life rafts include the following:

- Commercial, US Coast Guard (USCG) approved, 12 man
- Navy Mark 7

COMMERCIAL LIFE RAFTS

2-68. Commercial life rafts must be serviced at an approved servicing facility every 12 months. The servicing facilities must be approved by the life raft manufacturer, inspected by the Coast Guard, and issued a letter of approval by the USCG Commandant. Vessel masters should give careful attention to the selection of servicing facilities for inflatable life rafts. The painter must be connected to the ship by a weak link with a 500-pound breaking strength. Each inflatable life raft and container will have permanently attached a substantial nameplate of compatible material and which is embossed or imprinted with the name of the manufacturer. The nameplate must also have imprinted the approval number, the manufacturer's model and serial number, the number of persons for which the inflatable life raft is approved, lot number, the Marine Inspection Office identification letters, the date, and the letters "USCG". The raft container will also be provided with a stainless steel plate for showing a stamped record of the data of the annual inspections and the gas inflation tests described respectively. Particular attention should be given to liferaft launching arrangements when converting to commercial life rafts. In accordance with DA Pam 750-8, para 3-8, request a special mission modification. COMMERCIAL LIFE RAFTS ARE AUTHORIZED ONLY ON 900 SERIES SMALL TUG, forward deployed vessels, or other small vessels as designated by the Marine Safety Office. The Theater Support

Vessel uses a modified commercial marine evacuation system for rapid evacuation of large numbers of passengers.

LOCATION AND STOWAGE

2-69. Inflatable life raft stowage should be located to provide the following for each life raft:

- Stowed so that when the vessel sinks the survival craft floats free and, if inflatable, inflates automatically.
- Stowed in a position that is readily accessible to crewmembers for launching in less than 5 minutes, or provided with a remotely operated device that releases the life raft into launching position or into the water.
- Stowed in a way that permits manual release from its securing arrangements, without shifting in its mount.
- Ready for immediate use so that crewmembers can carry out preparations for embarkation and launching.
- Stowed in a way that neither the life raft nor its stowage arrangements will interfere with the embarkation and operation of any other life raft at any other launching station.
- Stowed in a way that any protective covers will not interfere with launching and embarkation.
- Stowed, as far as practicable, in a position sheltered from breaking seas and protected from damage by fire.

2-70. A mechanical, manually operated device to assist in launching a life raft must be provided if:

- The life raft weights more than 90.7 kilograms (200 pounds); and
- The life raft requires lifting more than 300 vertical millimeters (one vertical foot) to be launched.
- To permit ready manual overboard launching into the water without hitting obstructions.
- To be clear of overhead obstructions.
- To avoid the effects of heavy seas,
- To interfere as little as possible with normal shipboard activity.
- They shall be located, longitudinally, where they will provide the maximum practical distribution of lifesaving facilities. Furthermore, instructions shall be plainly visible and the station lighted by the vessels emergency lighting system.

SECURING

2-71. Equipment for securing the life rafts in their stowage mounts will include a hydrostatic release assembly that permits automatic and manual release. This provides for quick release of the life raft from its stowage for manual launching, or release from its stowage from hydrostatic pressure, resulting from a depth of 10 to 40 feet of seawater.

2-72. Commercial life rafts need to secure their sea painter with a float-free link (weak link) permanently attached to the vessel. The sea painter remains with the life raft.

2-73. The Navy MK 7 life raft sea painter should be tied or lashed to the ship in any manner. The Navy life raft sea painter will detach from the life raft upon completion of inflation.

SEA PAINTER

2-74. When manually launching the life raft, the sea painter will be attached directly to the ship structure adjacent to the stowage.

CAUTION

Rubber stoppers are inserted into the end of the sea painter line assembly to prevent entry of water into the life raft assembly and prevent accidental pay out of the painter line from its internal spool. If the stopper falls out, it is possible for high winds or seas to cause the entire 100' of spooled line to pay out, thus rendering the life raft inoperable, requiring its removal and transfer to a Navy repair facility for replacement of the sea painter spool. Properly installed rubber stoppers should be inserted until 1/8" to 1/16" of the stopper is visible uniformly about the circumference of the stopper on the outside of the life raft container. If there is ¼" or more of the rubber stopper exposed outside the life raft container, there is insufficient interference between the stopper and hole to prevent the stopper from coming loose in heavy weather. In no case shall the stopper be pushed flush or into the assembly.

MARKINGS

2-75. Markings on life rafts shall be clearly and legibly applied in a color contrasting to its background. Required markings are vessel hull number, "US ARMY", life raft number and capacity on the outer protective shell. Use materials that are permanent for the life of the inflatable life raft. Place instructions for launching and inflating the raft, righting the raft, and so on, in a conspicuous lighted area near the life raft. Recertification dates are located in the container handholds or under the black moisture band by the painter stopper. Stencil the vessel's hull number or name and the life raft number on the case.

NAVY LIFE RAFTS

NAVY MARK 7

2-76. The Mk-7 is constructed of polyurethane coated nylon fabric that is radio frequency welded making the Mk-7 very abrasion and puncture resistant with strong, durable seam construction. The periphery of the life raft bottom is arrayed with a series of weighted ballast bags that fill with water when the life raft is inflated providing stability to the life raft reducing the possibility that the raft can be capsized by wind and sea. An inflatable boarding ramp with webbing ladder projects from the life raft hull to provide easy entry from the water. In the event the life raft was to inflate upside down, a righting ladder is attached to the bottom so that one person can easily right the life raft. The Mk-7 is covered with a high visibility, double layered canopy, which can be secured, closed to protect the occupants from the weather. The canopy is equipped with a port so that a lookout can be placed to observe rescue craft with the canopy remaining closed. The canopy is also equipped with reflector tape to increase the visibility of the raft to rescue craft and a rain catchments system to collect rainwater to augment drinking water supplies. Life lines are located both inside and outside the life raft. The life raft floor is covered with an aluminized foam panel to provide the survivors thermal protection from cold sea water. The Mk-7 is equipped with a breathable air inflation system that can inflate the life raft in less than 30 seconds. The inflatable chambers have pressure relief valves to vent any excess air pressure and topping-off valves so that air can be added to maintain adequate air pressure and rigidity.

2-77. A sea anchor is deployed upon inflation to reduce drift of the life raft from the area. Once the life raft is inflated, flashing lights on the canopy are activated to allow survivors and rescue craft to locate the life raft if it is night time.

RECERTIFICATION AND REPAIRS

2-78. The Mk-7 life raft has a certified service period of five years. When the life raft reaches its certification expiration, it must be taken to a U.S. Navy approved life raft certification facility. Currently these are located in Portsmouth, VA, Mayport FL, San Diego, CA and Yokosuka, Japan. The U.S. Navy maintains a database that tracks all the Mk-7 life rafts. It keeps records on the life rafts' history, certifications, locations, conditions, and detailed information on vital components such as the inflation system. The database can be accessed by going to the Combatant Craft Department website, http://www.boats.dt.navy.mil/ and clicking on the button for the life raft database. Users must request an account to be able to access the database. Access to the database can help ships keep track of their life raft assets and manage the maintenance schedules for their life rafts.

2-79. Commercial life rafts have an annual recertification requirement plus a five year, in five year increments, (from date of manufacture) inflation test recertification. Any US Coast Guard approved commercial life raft maintenance facility can conduct recertification.

HYDROSTATIC RELEASE UNITS

2-80. This section contains information on operation and installation of the three types of Hydrostatic Release Units use on US watercraft. These units consist of:

- Release Lifesaving Unit
- Hydrostatic Release Unit, Navy Can
- Hammer H2O Hydrostatic Release Device

NOTE: Hydrostatic Release Units are also referred to as Hydraulic Release Devices or Units.

RELEASE LIFESAVING UNIT

2-81. The Hydrostatic Release Unit, Navy Can, NSN 4220-01-279-7287, has been discontinued. The replacement for the hydrostatic release is the Release Lifesaving Equipment. All existing hydrostatic releases will be replaced through attrition by normal supply acquisition process.

2-82. The Navy calls for the Release Lifesaving Equipment to be recertified every 5 years. This is the same time frame as for the life rafts. The following maintenance and inspection criteria, has been established for the Hydrostatic Lifesaving Equipment Release.

2-83. PMCS (Semi-Annual) includes:
- Inspect Diaphragm Type Hydrostatic Release Device Jaw Engagement Orientation.
- Ensure that both jaws of hydrostatic release are fully engaged. Full engagement is when inner surfaces of two jaws are in contact with larger diameter of plunger shaft.
- Inspect hydrostatic release device orientation to ensure that diaphragm housing and shorter arm are connected to deck and not retaining harness.
- Ensure that the hydrostatic release device is installed with the plunger facing inboard (towards traffic) and ensure that the release pin is installed to avoid inadvertent release.
-
- Inspect Diaphragm Type Hydrostatic Release Device.
- Inspect parts of diaphragm type release device(s).
- External surfaces (static ports) for paint or lubricant clogging. Paint or lubricant on external surfaces may clog static ports and prevent release device from operating.
- Connect linkage for corrosion, cracks, distortion and burrs
- Push button for clogging or distortions.
- Ensure safety pin remains attached
- Safety pin for corrosion or missing parts.
- Diaphragm halves mounting bolts for loose or missing bolts or nuts
- Inspect device for damage, distortion, corrosion or missing parts. Replace corroded, damaged or missing components or the entire unit as necessary.

HYDROSTATIC RELEASE UNIT, NAVY – "CAN"

NOTE: This device is to be replaced through attrition when the existing units become unserviceable.

2-84. This hydrostatic release unit (herein called the "CAN") consists of three components:
- aluminum can with a shield
- two end brackets
- hair pin/pull ring release mechanism.

2-85. The "CAN" has numerous advantages over the old diaphragm type. These advantages include the following:
- This device does not have an expiration date (which will reduce procurement cost, reduce maintenance, decrease inspection, and have no pressure test requirements).
- Components are replaceable off the shelf.
- The thin walled aluminum can is designed to be crushed by the pressure of water at a depth of 25 feet plus or minus 15 feet.
- The wall thickness of the can is such that it can be damaged during installation or inspection. A stainless steel shield is installed around the "CAN" to prevent this type of inadvertent damage as well as damage from wave slap or gun blast overpressure.
- The thin walled aluminum can forms a water tight seal that is the heart of the hydrostatic release. It is designed, if the ship sinks, to be crushed by the pressure of water; thus releasing the life raft. Collapsing of the aluminum hydrostatic can will allow the end brackets to release the life raft from its stowage.

> **WARNING**
>
> The "CAN" type hydrostatic release device does not require a pressure test. It is designed to collapse by water pressure. Pressure testing will destroy this device.

2-86. A stainless steel hair pin, a pull ring, and a safety sash chain constitute the components that provide for the manual release capabilities of this hydrostatic device. By removing the hair pin, the ring pin can be easily removed with the pull ring. The sash chain connects the hair pin and the pull ring to prevent losing these pieces during scheduled inspection and maintenance.

2-87. The "CAN" type hydrostatic release device is to be installed with the end bracket having the fixed pin and the pin retained by the bolt and nut attached to the life raft foundation of the vessel. The other end bracket with the fixed pin, the hair pin, and pull ring is attached to the retaining harness. The open end of the stainless steel shield around the can must face AFT to reduce the amount of water or air being trapped between the shield and the can. The hair pin should be installed in the end bracket facing the direction that would allow for easy removal when manually launching the raft. The retaining harness assembly (including the hydrostatic release device) is torque between 8- and 10-foot pounds to properly secure the raft in its stowage.

HAMMER H20 HYDROSTATIC RELEASE UNIT

2-88. The following is a description of the commercial Hammer H20 Hydrostatic Release unit.

2-89. The Hammer is a disposable, non-serviceable release unit. The Hammer H20 is simple to install and clearly marked with an expiration date (two years from installation). After installation, the unit remains in service with no periodic maintenance or service required. The Hammer does not rely on the positive buoyancy of the life raft to operate. It will release at all angles and needs only the required water pressure to activate.

2-90. The Hammer is simple in design. A loop of line is used to attach the retaining lashing for the raft to its cradle. This cord loop passes through the release mechanism. The glass-fiber, reinforced nylon casing accommodates a pressure chamber with a membrane activated spring loaded blade which cuts the line, releasing the life raft. The Hammer is designed to release the raft at a depth of 15 feet.

EMERGENCY ESCAPE BREATHING DEVICE (EEBD)

2-91. This section contains information about the Ocenco M-20.2 EEBD. This device is authorized for use onboard US Army watercraft.

> **WARNING**
>
> The EEBD is not to be confused as a fire-fighting device, such as the BA. It is to be strictly used as an emergency escape device. Improper use of this equipment may result in injury or death. Personnel should receive adequate training prior to use, including the limitations to which the equipment is subject. Personnel are reminded that there is no substitute for alertness, common sense, and self-discipline.

DESCRIPTION

2-92. Emergency Escape Breathing Device (EEBD) is a self-contained respiratory protection system in atmospheres containing toxic gas or in atmospheres that do not have enough oxygen. The Ocenco EEBD provides enough air for personnel to escape from below decks to the weather deck. It can be wall mounted (orange case) or belt-worn to provide quick and easy access in emergency situation. Ocenco EEBD unit is a phased replacement to the SCOTT EEBD unit. Ocenco EEBD unit provides air through a mouthpiece unlike the SCOTT EEBD, which provides air into a hood. The EEBD will also provide oxygen to trapped personnel awaiting rescue in contaminated atmospheres. The clear Teflon hood provides excellent heat and chemical protection. The Ocenco M-20.2 EEBD consists of the following:

- Nose Clip – the yellow nose clip is permanently attached to the mouthpiece. NOTE – NIOSH Requirement used in conjunction with mouthpiece.
- Oxygen Cylinder – stainless steel cylinder holds 100% medical grade oxygen. Holds 27 liters of oxygen.
- Oxygen Regulator – starts the flow of oxygen and regulates the oxygen flow during high work rates.
- Activation Cable – stainless steel activation cable is attached to the oxygen regulator and permanently attached to the inside of the clear base, once it is disconnected from the case, oxygen will flow immediately.
- Gauge – indicates the amounts of oxygen in the cylinder. The green zone indicates unit is ready for use. The red zone indicates the cylinder is low on oxygen and should be removed from service. NOTE – Two different configurations of gauges being supplied by Ocenco. One is dial type and the other is coil type.
- Breathing Bag – an air reservoir that receives oxygen from the oxygen regulator and exhaled air from the scrubber. NOTE – Two different colors breathing bags available: Black and Tan, Kevlar type.
- Relief Valve – a one-way valve automatically allows excess air in the bag to vent
- Face Shield – optional

NOTE: The optional face shield is to help protect face and eyes from smoke or chemical vapor.

2-93. The EEBD is rated by National Institute for Occupational Safety and Health (NIOSH) for a minimum of ten minutes of operation. NOTE – Testing has shown that unit can last between 15 and 20 minutes depending on breathing rate and up to 32 minutes, if user is trapped and waiting rescue. The unit can be donned and activated in 10 seconds. Activation is accomplished by pulling the unit from its case, which will automatically start oxygen flow. Once activated, the unit cannot be shut off. It is a single use, disposable unit.

2-94. The service life of an EEBD is 15 years. It can be belt-worn continuously for 5 years in engineering machinery spaces or confined spaces before returning to a wall mounted bracket for the remainder of its service life. The weight of the breathing apparatus is 1.98 lb; the apparatus with case 2.87 lb. The maximum operating temperature is 140 degrees F and the minimum is 10 degrees F.

OPERATING PROCEDURES

2-95. The following instructions will enable users to familiarize themselves with those procedures which are necessary to operate the escape device in actual in-service conditions. These instructions will also describe the capabilities and limitations of the equipment in so far as the user is concerned.

NOTE: When donning from orange stowage case, place unit on ground with orange pull-tab away from body. Open orange stowage case, move hand over unit lift yellow lever then remove unit and discard cover case and follow steps 3-7.

- Lift yellow lever and discard cover case.
- Remove unit by pulling on yellow neck strap upward, starts oxygen flow.

- Put yellow neck strap over head.
- Insert yellow mouthpiece into mouth.

NOTE: Make sure the nose clip is flipped out before inserting yellow mouthpiece into mouth to ensure that wearer of unit does not breathe through nose while in toxic atmospheres.

- Fit yellow nose clip.
- Inhale through mouth and escape.
- Fit and adjust yellow neck strap and face shield by pulling on O-rings for a secure fit.

NOTE: Face Shield is optional, to protect faces and eyes from smoke and fire.

REMOVING AND DISPOSAL

2-96. Once the wearer is clear of the hazardous area the following steps indicate how to properly remove and disposition the device.
- Loosen neck strap.
- Pull forward overhead.
- Remove unit.
- Turn into Unit Hazmat for disposal.

MAINTENANCE

2-97. Maintenance will be conducted IAW Army Regulations and manufactures guidelines. The EEBD, whether wall mounted or belt-worn, should be inspected for any indications of high force impact. Check to make sure unit does not have:
- case cracks
- burns
- deformities
- excessively worn parts
- damaged latch or cover band
- bent gauge
- broken indicator needle
- dirt, debris or moisture visible through gauge window
- broken belt loops
- missing tamper indicating ball

2-98. If any of these indications of high force impact are observed, or if the pressure gauge is out of the green zone, remove unit from service. To ensure top condition, always provide best possible care and follow PMCS procedures.

SERVICE LIFE

2-99. The Ocenco EEBD has a service life of 15 years from the date of manufacture, provided the conditions of use are observed. The Ocenco EEBD must be either stored in the orange stowage container or belt worn. The Ocenco EEBD may be deployed in the belt worn configuration for 5 continuous years during its 15 year service life.

TRAINING EEBD

2-100.   The training EEBD is light blue and comes in a light blue stowage case. It comes equipped with 3 mouthpieces. It does not come equipped with a Teflon hood, 10-minute supply of oxygen, or a lithium hydroxide scrubber.

> To order the training video from Ocenco:
> Telephone: (262) 947-9000, FAX: (262) 947-9020
> E-mail: eebd@ocenco.com, web site: www.ocenco.com

NOTE: EEBD cases should be marked with vessel name or number.

WARNING

The EEBD is a one-time use throwaway device.

NOTE: EEBD canisters must be disposed of IAW HAZMAT/Environmental SOP.

STOKES LITTER

2-101.   This section contains information about the corrosion-resistant steel litter used onboard Army watercraft.

CONFIGURATION

2-102.   Stokes litters shall be configured for their intended application and shall not be used otherwise.
- Ashore - Stokes litters used for transporting a person for land operations require no modifications. Steel or aluminum litters may be used.
- Over Water - Stokes litters used for transporting a person onboard boats, over the water, or retrieving a person overboard, shall be configured with a flotation kit assembly (includes tow flotation tubes with covers, one chest pad with cover, five restraint straps, and one ballast bar), slat set, hoisting slings, and tending lines.
- Hoisting - Stokes litters intended for shipboard or helicopter-hoisting operations (using the ship's or aircraft's hoist) shall be equipped with the standard hoisting sling.

WARNING

Only steel litters are authorized for hoisting operations.

NOTE: Class "A" Army watercraft shall maintain at least one corrosion resistant steel Stokes litter rigged for over water use. Class "B" and "C" Army watercraft may maintain at least one corrosion-resistant steel Stokes litter rigged for over water use.

MODIFICATIONS

2-103.   The following paragraphs describe modifications to the Stokes litter.

2-104. **Flotation Tubes, Chest Pad, and Ballast Weight.** To attach the flotation tubes, chest pad, and ballast weight, complete the following steps:

- Route one end of the flotation tube webbing tie over the top (¾-inch) litter tube and the other end of webbing tie under the lower (3/8-inch) litter tube.
- Be sure to position ties over the outside of the flotation tube and in the location illustrated in Figure.
- Tie or connect ends of webbing together using a square knot or buckles (if so equipped). Tack free ends of webbing using 6-cord nylon thread.
- Route chest pad strap through retainer straps on cover and attach to the lower (3/8-inch) litter tube as illustrated in Figure.
- To make litter self-righting, attach a 12-pound ballast bar to foot of litter.

WARNING

It is essential that the two flotation tubes, chest pad, and ballast bar be positioned at the precise locations on the litter as noted in instructions. If the tubes are positioned too high or too low, the litter may not right itself or keep the patient's head above water.

2-105. **Hoisting Sling**. Attach the sling.

WARNING

Use two swagging sleeves on each end of the hoisting sling when attaching it to the litter tube.

2-106. **Restraint Straps.** Attach four restraint straps. To attach straps to the 3/8-inch tube, pass loop end of restraint strap around outside and under tube, passing strap between wire mesh and tube. Pass opposite end through loop and pull strap tight.

2-107. **Tending Lines.** Stokes litters shall have tending lines installed so the litter can be held in position and recovered from alongside a vessel for rescue from the water. Use manila line of sufficient length to allow lowering of litter to the water. Attach tending lines using an eye splice, to the ¾-inch tubes at stations 3 and 6 of litter.

INSPECTION

2-108. Stokes litters and associated equipment shall be inspected after each use but not less than once every three (3) months. The latest date of inspection and proof test shall be stenciled on the bottom of the slat set (in trunk section of litter). The stencil shall be of ½-inch letters.

2-109. The following paragraphs contain requirements for inspection of the litter, flotation equipment, hoisting sling, and tending lines.

2-110. **Litter.** Inspect litter for cracked welds, cracked tubes, rust, pinholes, security and condition of wire mesh, and evidence of wear on the sling attachment points. Inspect restraint straps for security, condition, and quantity (minimum of four per litter).

2-111. **Flotation Equipment.** After use in salt water, flotation equipment shall be rinsed in fresh water and dried before storage. Flotation equipment shall be thoroughly inspected for wear, rotting, mildew, mold, tears, cuts, broken stitches, and frayed fabric.

2-112. **Hoisting Sling.** The hoisting sling shall be inspected for corrosion, fraying, or deterioration. The hoisting requires test loaded certification every 6 months.

2-113. **Tending Lines.** Inspect manila tending lines for condition and security. Lines that are frayed or show signs of weathering or rot shall be replaced.

PROOF TESTING

2-114. Litters equipped with a hoisting sling shall be proof tested every 6 months. To proof test litter, complete the following steps:

- Distribute 400 pounds evenly in the litter and hoist clear of the deck.
- With litter suspended, inspect litter and sling for deformities.
- Inspect sling for even load distribution at all attachment points.

MAINTENANCE

2-115. Maintenance of the litter, hoisting slings, flotation equipment, and chest pad consists of minor repairs, replacement, and cleaning. Repairs for aluminum litters are limited to removal of surface corrosion and application of primer to rework areas. Cracked welds or cracked tube members are cause for replacement. To maintain equipment, complete the following steps:

WARNINGS

No weld repairs shall be attempted on aluminum litters. Aluminum litters shall be marked "NOT TO BE USED FOR HOISTING OR HIGH-LINE OPERATIONS."

- Weld repairs for steel litters are permitted using heliarc method only. After a weld repair, litter shall be proof tested as described in paragraph.
- Replace hoisting slings that show signs of corrosion, fraying, or deterioration.
- After each use in salt water, remove flotation collar and chest pad from the litter, rinse in fresh water, and dry before reinstallation.

HEAVING / SAFETY LINE

2-116. This section contains information about the heaving/safety line. The heaving/safety line is authorized aboard US Army watercraft. The heaving/safety line is a two-part item. One part is the weighted ball and the other part is the line itself. See Figure 2-16.

Figure 2-15, Heaving Safety Line

DESCRIPTION

2-117. The weighted ball and the line are described below.

- **Ball, Heaving Line.** The ball is fluorescent orange in color and is approximately 4 ¼ inches in diameter. The ball weighs approximately 10 ounces with a 1/4-inch wall thickness. The ball has two

1/2-inch holes in diameter going through the center which is injection molded airtight into the ball on 1 1/2-inch centers. It is extremely strong and durable and soft to the touch as not to harm anyone while being thrown. It will float with approximately 60 percent of its body out of water.

- **Line, Heaving/Safety.** The line is made of polypropylene materials with ultraviolet additives. It is called 3/8-inch dual braid, and is woven with an inner and outer core and an outer braid to prevent tangling. The line is soft to the touch, but very strong with an approximately 1500# breaking strength and is 100 percent floatable. The ball may be used as a fender for small boats. It can also be used to retrieve a person from overboard by sliding the line through the holes in the ball to make a loop.

LINE THROWING DEVICE

2-118. This section will cover the requirement, stowage, and safety precautions pertaining to the throwing device used aboard army watercraft.

SHOULDER GUN EQUIPMENT

2-119. The shoulder gun type (herein also called the US "appliance") used aboard US Army watercraft is required to have the following equipment carried for each shoulder gun:

- Ten service projectiles.
- Twenty-five cartridges.
- Four service lines.
- One cleaning rod with brush, one can of oil, and 12 wiping patches.
- One set of instructions from the manufacturer.
- One auxiliary line that is made of one of the following:
  - Manila and is at least 500 feet long and 3 inches or more in circumference; or A synthetic material and is at least 500 feet long, is certified by the manufacturer to have a minimum breaking strength of 9,000 pounds, and inhibited to resist the effects of ultraviolet rays.
  - In the event that a vessel is deployed forward of its home base, a SOLAS standard line-throwing device can be procured and utilized when items within the DoD procurement system cannot be obtained.

NOTE: All equipment must be stowed with the appliance in a box or case, except for the service lines and the auxiliary line, which may be stowed in an accessible location nearby.

DRILLS

2-120. It is the duty of the vessel master to drill the crew in the use of the appliance. The device is to be exercised at least once every quarter. The service line will not be used for drill purposes. Test firing may be accomplished using the regular cartridge and projectile with any flexible line of proper size and lengths, suitably faked or laid out. There are no replacement cartridges available within the supply system so due care is required to ensure vessel retains adequate supply on board. Vessels must log exercise of all safety equipment IAW para 1-29 of this manual.

ACCESSIBILITY

2-121. The line throwing appliance and its equipment will be kept easily and readily accessible and ready for use. No part of this equipment will be used for any other purpose. When firing the appliance, the operating instructions and safety precautions furnished by the manufacturer shall be followed.

WARNING

As with any firearm or firing mechanism, the appliance should be handled with caution and only when its use is required. Misuse of this piece of equipment could cause personal injury or death.

NEIL ROBERTSON STRETCHER

2-122. This stretcher is designed for removing an injured person from engine room spaces, holds, and other compartments. This stretcher is used where access hatches are too small and through narrow passageways and ladder wells to permit the use of regular stretchers or litters. This stretcher is authorized for use on US Army watercraft.

DESCRIPTION

2-123. The Neil Robertson stretcher is made of semi-rigid canvas. When firmly wrapped (mummy-fashioned) around the victim (see figure 2-17), it gives sufficient support so the victim may be lifted vertically. To keep the injured person from swaying against bulkheads and hatchways while being lifted, a guideline is tied to the victim's ankles.

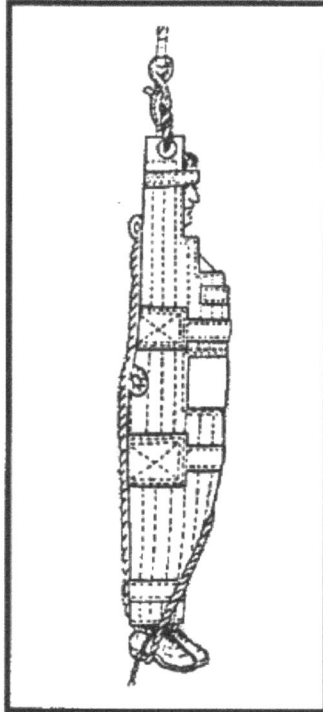

Figure 2-16. Neil Robertson stretcher

STOWAGE

2-124. Stretchers of this type should be kept onboard in appropriate places ready for use.

RESCUE / WORK BOAT

STOWAGE

- Rescue boats must be stowed:
- To be ready for launching in not more than 5 minutes.
- In a position suitable for launching and recovery;
- In a way that neither the rescue boat nor its stowage arrangements will interfere with the operation of any survival craft at any other launching station;

2-125. Each rescue boat must have a means provided for recharging the rescue boat batteries from the vessel's power supply at a supply voltage not exceeding 50 volts.

2-126. Each inflated rescue boat must be kept fully inflated at all times.

RESCUE BOAT EQUIPMENT

- Paddles, bailer onboard in operational condition
- PFDs onboard and serviceable
- Navigation lights operational
- Becketed lifelines installed both inside and outside of the boat
- Lighted magnetic compass onboard
- Sea anchor and hawser
- Painter and release device
- 50 meter buoyant line
- Waterproof flashlight, w/kit (spare batteries/ bulb in w/t case)
- Horn
- Waterproof First Aid kit
- Searchlight
- Buoyant safety knife
- Manual pump/bellows
- Repair kit
- Safety boat hook
- Emergency signal kit
- Exposure suits suitable for entire
- Radio tested

All rescue boat equipment not integral with the hull will be marked "U.S. Army" and with the vessel name/number.

LAUNCHING AND RECOVERY ARRANGEMENTS

2-127. Each rescue boat must be capable of being launched with the vessel making headway of five (5) knots in calm water. A painter may be used to meet this requirement. Each rescue boat embarkation and launching arrangement must permit the rescue boat to be boarded and launched in the shortest time possible. If the launching arrangement uses a single fall, the rescue boat may have an automatic disengaging apparatus release mechanism. Rapid recovery of the rescue boat must be possible when loaded with its full complement of persons and equipment. Each rescue boat launching appliance must be fitted with a powered winch motor. Each rescue boat launching appliance must be capable of hoisting the rescue boat when loaded with its full rescue boat complement of persons and equipment at a rate of not less than 0.25 meters per second (0.5 knots or 50 feet per minute).

MARKINGS ON RESCUE BOATS

2-128. Each rescue boat must be plainly marked as follows:

- Each side of each rescue boat bow must be marked in block capital letters and numbers with the name of the vessel and hull number
- The number of persons for which the boat is equipped must be clearly marked, preferably on the bow, in permanent markings

RESCUE BOAT EQUIPMENT

- Paddles, bailer onboard in operational condition
- PFDs onboard and serviceable
- Navigation lights operational
- Becketed lifelines installed both inside and outside of the boat
- Lighted magnetic compass onboard
- Sea anchor and hawser
- Painter and release device
- 50 meter buoyant line
- Waterproof flashlight, w/kit (spare batteries/ bulb in w/t case)
- Horn
- Waterproof First Aid kit
- Searchlight
- Buoyant safety knife
- Manual pump/bellows
- Repair kit
- Safety boat hook
- Emergency signal kit
- Exposure suits suitable for entire
- Radio tested

All rescue boat equipment not integral with the hull will be marked "U.S. Army" and with the vessel name/number.

LAUNCHING AND RECOVERY ARRANGEMENTS

2-127. Each rescue boat must be capable of being launched with the vessel making headway of five (5) knots in calm water. A painter may be used to meet this requirement. Each rescue boat embarkation and launching arrangement must permit the rescue boat to be boarded and launched in the shortest time possible. If the launching arrangement uses a single fall, the rescue boat may have an automatic disengaging apparatus release mechanism. Rapid recovery of the rescue boat must be possible when loaded with its full complement of persons and equipment. Each rescue boat launching appliance must be fitted with a powered winch motor. Each rescue boat launching appliance must be capable of hoisting the rescue boat when loaded with its full rescue boat complement of persons and equipment at a rate of not less than 0.25 meters per second (0.5 knots or 50 feet per minute).

MARKINGS ON RESCUE BOATS

2-128. Each rescue boat must be plainly marked as follows:
- Each side of each rescue boat bow must be marked in block capital letters and numbers with the name of the vessel and hull number
- The number of persons for which the boat is equipped must be clearly marked, preferably on the bow, in permanent markings

Chapter 3

FIRE FIGHTING

CHEMISTRY OF FIRE

3-1. **Start of a fire.** Matter exists in one of three states: solid, liquid, and gas (vapor). For a solid or liquid fuel to burn enough heat must be applied for vapors to form on the surface. These vapors must intermix with the oxygen in the surrounding air in order for a flammable mixture to form.

3-2. **Fire triangle.** Fuel, oxygen, and heat are required for combustion. The fire triangle represents these requirements. The three sides of the fire triangle represent fuel, oxygen and heat. If any side of the triangle is missing, a fire cannot start. If any side of the fire triangle is removed the fire will go out. The ignition temperature of a substance is the lowest temperature at which sustained combustion will occur without the application of a spark or flame. The flashpoint is the lowest temperature at which a liquid gives off sufficient vapor to form an ignitable mixture. Sustained combustion takes place at a slightly higher temperature above flashpoint (referred to as the fire point of a liquid).

3-3. **Fire Tetrahedron.** The Fire Tetrahedron is a better representation of the combustion process. The four sides of the tetrahedron represent: fuel, oxygen, heat and the chain reaction. Combustion is sustained through a chemical chain reaction. Disrupting the chain reaction is a means of extinguishing a fire.

3-4. **Oxidation.** A chemical process in which a substance combines with oxygen. Rusting iron/rotting wood are examples of slow oxidation. Fire (or combustion) is an example of rapid oxidation. During oxidation, energy is given off, usually in the form of flames. When a substance combines with oxygen at a very high rate, the energy given off as heat and light is so rapid that we can feel the heat and see the light in the form of flames.

3-5. **Combustion.** Combustion is the rapid oxidation of millions of vapor molecules. During combustion, radiant heat is released, which radiates in all directions. Heat that radiates back to the fuel is called radiation feedback. The radiation feedback creates more vapors from the fuel source. A self sustaining reaction starts, burning vapor produces heat which releases and ignites more vapor. This chain reaction will continue as long as fuel, oxygen and sufficient heat are generated to create more vapors and raise the vapors to the fuels ignition temperature.

3-6. **Heat transfer.** Heat transfer occurs through one or more of three different modes: Conduction, Radiation and Convection. Conduction is the transfer of heat through a solid body. Radiation is the transfer of heat across an intervening space; no material substances involved. Convection is the transfer of heat through the motion of circulating gases or liquids.

3-7. **Hazardous products of combustion.** Flames and heat are obvious hazards of any fire, yet gases generated by combustion are also lethal. Carbon monoxide is an abundant byproduct of combustion, resulting from incomplete combustion. Carbon dioxide is also an abundant byproduct, resulting from complete combustion. Carbon monoxide is the more dangerous of the two. The presence of byproduct gases reduces the oxygen content of the surrounding air, which is normally 21 percent. Carbon monoxide works on the respiratory system. Above normal CO concentrations in the air reduces the amount of oxygen that is absorbed in the lungs, impairing muscular control. Smoke generated by a fire creates an atmosphere that reduces visibility and impairs breathing.

THEORY OF EXTINGUISHMENTS

3-8. A fire can be extinguished by the following methods:

- **Removing the fuel.** Removing the fuel source from a fire will extinguish the fire.
- **Removing the oxygen.** A fire can be extinguished by reducing the oxygen percentage in the air. Lowering the oxygen content in the air will decrease the intensity of the fire, eventually extinguishing the fire depending on the properties of the fuel. Fires can smolder in as low as 6% oxygen content.
- **Removing the heat.** Water, applied as a low velocity fog is the most effective means of removing heat from ordinary combustible material
- **Break the chain reaction.** A fire can be extinguished by disrupting the chemical process that sustains the fire.

3-9. For Firefighting purposes, there are six classes of fire; Class A, Class B, Class AB, Class AC, Class BC, Class D.

- Class A fires involve wood and wood products, cloth, textiles and fibrous materials, paper and paper products and are extinguished with water. High velocity fog is the preferred method of extinguishing a class A fire.
- Class B fires involve gasoline, solvents, oil, and other flammable liquids. These fires are extinguished with Water fog, AFFF, FM 200, or CO_2. Flammable gas fires should never be extinguished unless there is a reasonable certainty that the flow of gas can be secured. An explosive condition can result, which can be a greater hazard than the fire.
- Class AB fires involve solid fuel combined with liquid or gaseous fuel. These fires are fought as a class B fire.
- Class AC and BC fires are energized electrical fires that are attacked using nonconductive agents such as CO_2 or FM 200 at prescribed distances. The most effective tactic is to de-energize and handle the fire as a class A or B fire.
- Class D fires involve combustible metals such as magnesium and titanium. Dry powders are the only agent used for extinguishing Class D fires. DO NOT use water on a Class D fire, water will increase the intensity of the fire and spread it. Do not confuse dry powder with dry chemical.

FIRE FIGHTING AGENTS

WATER

3-10. There are many materials that may be used as fire fighting agents. Water is used primarily as a cooling agent. If the surface temperature of a fire can be lowered below the fuel's ignition point, the fire will be extinguished. A secondary method of water extinguishment is caused by steam smothering. Army watercrafts are equipped with both the Navy's All-Purpose Nozzle (APN) and the Navy's Vari-Nozzle for this purpose.

Fire Hose Nozzles

3-11. The Navy's All-Purpose Nozzle is used on internal fire hoses to deliver water for three different applications:

- In the form of straight stream, the nozzle provides maximum velocity and range, but has little cooling affect. Straight stream should be used during the overhaul of fires only.
- In the form of high velocity water fog, the nozzle provides a cone-shaped fixed pattern of small water particles. This is the best pattern for combating fire.
- In the form of low velocity water fog, the nozzle is equipped with an applicator that provides fine water particles in a wide umbrella-shaped cloud pattern. This pattern has the greatest cooling effect and should be used for personnel protection.

3-12. Applicators are tubes or pipes angled at 60 or 90 degrees.

- 1 ½ nozzles use a 4 ft. applicator with a 60° angle and a 10 ft. applicator with a 90° angle.

3-13. These applicators are used to reach around corners, through narrow openings, and to cool the primary fire fighting hose team.

3-14. The Navy Vari-Nozzle is used on exterior fire hoses and can deliver water in the form of a straight stream, high velocity fog or low velocity fog. The nozzle tip is rotated to deliver the various patterns.

FOAM

3-15. There are two basic types of foam: chemical and mechanical. Due to recent legal actions foam can only be utilized during real emergencies on board Army Watercraft. Crew training using foam must be conducted in a controlled, land-based element following DOD policies and EPA standards. Foam is found on vessels that have manned machinery spaces, carry large capacity petroleum tanks, or tow fuel barges great distances. Foam is carried either in 5 gallon sealed plastic containers located in the manned machinery spaces or in large holding tanks. The shelf life of foam is 25 years however should the foam become contaminated or diluted then it must be replaced. Foam stored in tanks must be tested every 2 years while foam in 5 gallon plastic containers is serviceable until exposed to the elements. Army Watercraft use the 6% concentrate foam in either gel or liquid form. Foams can be used on class A and B fires.

3-16. Chemical foam is formed by mixing an alkali (usually sodium bicarbonate) with an acid (usually aluminum sulfate) in water. Chemical foam has been phased out for shipboard use in favor of mechanical foam. Mechanical foam is produced by mixing foam concentrate with water to produce a foam solution. This foam solution is then mixed with air creating finished foam. The bubbles are formed by the turbulent mixing of air and the foam solution. Aqueous Film Forming Foam (AFFF) provides three fire extinguishing advantages:

- An aqueous film is formed on the surface of the fuel which prevents the escape of hydrocarbon fuel vapors.
- The layer of foam effectively excludes oxygen from the fuel surface.
- The water content of the foam provides a cooling effect.

3-17. Limitations of foam:

- Electrically conductive and should not be used on live electrical equipment.
- Should not be used on combustible metal fires.
- Should not be used on fires involving gases.

FIXED FIRE FIGHTING SYSTEMS

FM 200 / HEPTAFLOUROPROPANE (HFC-227EA)

3-18. FM 200 is the most common fixed fire fighting system on Army Watercraft. FM 200 is the brand name for heptaflouropropane (HFC-227EA). HydroFlouroCarbons (HFC) are composed of carbon, fluorine and hydrogen atoms. FM 200 extinguishes fire through heat absorption and to a lesser extent, chain breaking. The class of compound FM 200 belongs to are HFCs. HFCs are used in refrigeration and are a very effective heat transfer agent. HFCs remove heat energy from a fire so the fire cannot sustain itself. FM 200 also releases small amounts of free radicals upon exposure to flames. Free radicals inhibit the chain reaction responsible for combustion.

3-19. FM 200 will not damage delicate equipment because it does not have particulates or oily residues. It does not significantly reduce oxygen levels when deployed, making it safer for people. FM 200, when exposed to temperatures in excess of 1300 degrees F, will break down chemically and create Hydrogen Fluoride (HF) gas. . This system must be recertified annually by a manufacture's certified technician. The bottles have a shelf life of twelve (12) years between hydrostatic testing unless they have been discharged.

DANGER

HF gas is very toxic to humans. Concentrations above three (3) parts per million (PPM) can be harmful and/or fatal

3-20. Benefits of Using FM 200:

- Fast-Acting FM 200 can stop fires in just seconds.
- Extinguishing fires quickly means less damage, repair costs and extra safety.
- FM 200 has been tested extensively to ensure safe exposure to humans.
- FM 200 does not leave oily residues, particulates, water, or corrosive material. This eliminates collateral damage to delicate equipment.
- FM 200 has a low environmental impact because it has a low atmospheric lifetime. It also has zero potential to deplete the ozone layer.
- Small Space Requirement compared to other fire suppression systems, such as CO2 and inert gases require as much as seven times more storage space.
- Globally Accepted FM 200 is the most widely accepted clean agent in the world. It is used in many fire suppression systems.

3-21. HCF-227EA (Trade Name FM 200). The systems are fixed, total flooding HFC-227ea fire extinguishing systems (figure 3-1), with a general arrangement of actuation station, cylinder storage, alarms, and spaces protected by individual cylinder banks. Some exceptions are:

- The flammable liquid storage space.
- Provide new systems to be installed for the Logistics Support Vessel (LSV) flammable liquid storage spaces.

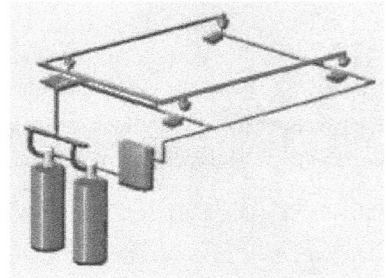

Figure 3-1, HFC-227EA (FM 200) Fire Extinguishing Systems

WHEN ALARM SOUNDS, VACATE AND SECURE THE SPACE TO KEEP THE FM-200 FROM ESCAPING OUT OF THE COMPARTMENT

3-22. The control cabinets or spaces containing valves or manifolds will be conspicuously identified in red letters at least 2 inches high. Example: "HFC-227EA or trade name (FM-200) FIRE APPARATUS."

3-23. FM-200 systems utilize CO2 gas to operate the time delay and alarm portion of the fire fighting system. The time delay and alarms allow up to 60 seconds for individuals in the spaces to escape before the main

3-23. FM-200 systems utilize CO2 gas to operate the time delay and alarm portion of the fire fighting system. The time delay and alarms allow up to 60 seconds for individuals in the spaces to escape before the main extinguishing agent is discharged. However, there is sufficient CO2 gas in the time delay and alarm operating system to cause breathing problems for anyone in the space when the devices are activated. Care needs to be taken to ensure that all individuals safely evacuate the space.

WATER WASHDOWN SYSTEM (WWS)

3-24. The Water Washdown System can be utilized separately as a stand-alone system but it is integral to the FM 200 system. Only occupied machinery spaces have this system added to the FM 200 system. It is part of the Fire Main system but is activated at locations outside the space being energized. This system must be activated prior to activating the FM 200 system and must continue until a minimum of 15 minutes AFTER the FM 200 bottles are emptied. Its primary purpose is to ensure that the space temperature remains below 1300 degrees F in order to minimize the creation of HF gas. As a stand-alone system it can be energized repeatedly as a fire fighting device.

CARBON DIOXIDE (CO2)

3-25. The fixed CO2 system is sized to the machinery space that it is designed to protect. It can be activated from one of two locations: locally at the bottles and external to the space being protected. This system must be recertified annually by a manufacture's certified technician. The bottles have a shelf life of 12 years between hydrostatic testing unless they have been discharged. Due to the oxygen displacement characteristics of this gas, there is a time delay of 60 seconds from the instant controls are activated until the actual gas is released. This time delay allows the space occupants to leave before the gas is released.

NOTE: Carbon Dioxide (CO2) is a dry, non-corrosive gas, which is inert when in contact with most substances, and will not leave a residue to damage machinery and electrical equipment.CO2 has limited cooling capabilities. The firefighter should have backup extinguishers

CO2 extinguishes the fire by diluting and displacing its oxygen supply. Individuals employing CO2 systems need to be aware of the dangerous situation caused by using CO2 in closed spaces.

PORTABLE FIRE EXTINGUISHERS

3-26. Portable fire extinguishers are used for fast attack on small fires. They can be carried to the fire easily, but have a very limited supply, in many cases only one minute or less. With that in mind, it is important to back up a portable fire extinguisher with a fire hose.

NOTE: Portable dry chemical fire extinguishers must be recertified every 6 years while CO2 fire extinguishers must be hydrostatically tested every 5 years from the manufactured date.

Class of Fire Extinguishers

3-27. Portable extinguishers are classified in two ways, with one or more letters and with a numeral. The letter or letters indicate the classes of fire that the extinguisher may be used on (i.e. A, B, C, or D). The numerals indicate the relative efficiency of the extinguisher or its size.

NOTE: National Fire Protection Association (NFPA) rates extinguisher efficiency with Arabic numbers. Underwriters Laboratory (UL) tests extinguishers on controlled fires to determine their NFPA ratings. Coast Guard uses Roman numerals to indicate the size of portable extinguishers.

CO2 Extinguisher

3-28. Carbon dioxide extinguishers are used primarily on class B, AC, and BC fires. The most common size of portable extinguishers contain from 5 to 20 pounds of CO2, not including the weight of the shell. Range varies between 3 to 8 feet, and the duration between 8 to 30 seconds, depending on the size. Their use will be confined to class B pool fires no greater than four square feet. These bottles must be hydrostatically tested every 5 years from manufacture date. The manufacture date can be found in one of three locations: on the label, stamped into the bottle bottom, or stamped into the bottle neck.

3-29. Prior to operating the cylinder, the operator must ground it to the deck to avoid shock from static electricity. In addition, the operator must grasp the hose handle and not the discharge horn, as the horn gets cold enough to frost over and cause severe frostbite. They should be stowed at temperatures below 130F to keep their internal pressure at a safe level.

ABC Extinguisher

3-30. Monoammonium phosphate (ABC, multi-purpose) dry chemical may, as its name implies, be used on Class A, B, and C fires and combinations of these. However, ABC dry chemical may only control, but not extinguish, some deep-seated Class A fires and an auxiliary extinguishment method, such as a water hose line, is required. These bottles must be hydrostatically tested every 6 years from manufacture date. The manufacture date can be found in one of three locations: on the label, stamped into the bottle bottom, or stamped into the bottle neck. All dry chemical agents may be used to extinguish fires involving the following.

- Flammable oils and gasses.
- Electrical equipment.
- Hoods, ducts, and cooking ranges in galleys.
- Machinery spaces, engine rooms, paint lockers, and tool lockers.

3-31. Dry chemical extinguishing agents are very effective on gas fires. However, gas flames should not be extinguished until the supply of fuel has been secured.

3-32. There are limitations on the use of dry chemical extinguishers. The following should be considered before use:
- The discharge of large amounts of dry chemical could affect people in the vicinity.
- Dry chemical may deposit an insulating coating on electronic or telephonic equipment, affecting the operation of the equipment.
- Dry chemicals are not effective on combustible metals such as magnesium, potassium, sodium, and their alloys, and in some cases may cause a violent reaction.
- Where moisture is present, dry chemical agent may corrode or stain surfaces on which it settles.

GALLEY FIRE SUPRESSION SYSTEM

3-33. The galley fire suppression system is proven and reliable. It is always there, and will act as designed when needed. However, there are fundamental actions that can be accomplished that will prevent the need for its use. There are different types of systems on Army Watercraft: fire suppression with nozzles and ventilation blocking which prevents the fire from spreading through the ventilation systems. These actions will help prevent a Galley fire:
- Keep all galley equipment clean and free of grease buildup.
- Never use flammable solvents or cleaners on the galley equipment. Flammable residues could be left behind and could ignite during subsequent use of the galley equipment.
- Operate the galley exhaust system when the heat producing equipment is preheating, cooking, or cooling after use. This helps to prevent excessive heat buildup which could cause unnecessary system actuation.
- Never operate the hood without the filters in place. Excessive grease could build up in the hood and ductwork. Clean the filters regularly.

- Never restrict the air intake passages. This reduces the efficiency of the hood exhaust system and could lead to excessive heat buildup and accumulation of fumes.
- Properly operate the grease extractors to ensure effective grease removal from the hood and duct system.
- Never tamper with the fire suppression system components such as the heat detectors, nozzles, agent storage canister, cables, or the fusible links.
- Report any damage or suspect component to the Vessel Master and Chief Engineer or the Assistant Chief Engineer immediately.
- Ensure that portable fire extinguishers are properly placed, installed, inspected, and available for use.

NOTE: U.S. Army vessels use a variety of Galley Equipment Fire Suppression Systems. Care should be taken to reference vessel TM's and Manufacturer's documentation with regard to the specific system installed on board your assigned vessel. Maintenance and operation procedures will vary between systems and due care must be taken to ensure that proper procedures are followed for each system.

REPORTING A FIRE

3-34. The crew member who discovers the fire or the indication of fire must sound the alarm promptly.

> **THE FIRE SIGNAL IS A CONTINUOUS BLAST OF THE SHIP'S WHISTLE FOR NOT LESS THAN 10 SECONDS SUPPLEMENTED BY THE CONTINUOUS RINGING OF THE GENERAL ALARM BELLS FOR NOT LESS THAN 10 SECONDS.**

3-35. No crew member should ever attempt to fight a fire, however small it may seem, until the alarm has been passed to the bridge.

3-36. The crew member who sounds the alarm must be sure to give the exact location of the fire, what class of fire and as much information as is available when passing word to the bridge.

3-37. Before a compartment or bulkhead door is opened to check for fire, the door should be examined. Look for the following:

- Discolored or blistered paint indicates fire behind the door or smoke puffing from cracks at the door seal.
- The bulkhead or door should be checked for heat with the back of the hand, touching the door is not necessary to feel the heat.
- Once a hidden fire has been located, the door to the area should not be opened until directed by the On-scene leader. Cool the door with water, if necessary, before opening.
- Always open the door from a position clear of the opening and opposite the hinges.

FIRE DRILLS AND TRAINING

3-38. The best organization and equipment is useless without trained personnel. Properly drilled crewmembers will lessen confusion during fires, increase the probability of proper initial actions taken against a fire, and enhance the predictability of fire fighting responses and uses. Vital to the effort, however, is continuity of personnel. That is, people assigned to the fire fighting party should retain that position even if other shipboard duties change. All members of a fire party should be cross trained for at least one other position on the fire party in order to provide frequent rotation. Ideally, everyone on the ship should be training to serve on a fire party since they may be needed to fight a major fire.

NOTE: It is also important to use the drill as an opportunity to test, inspect, and repair all fire fighting equipment. Each fire drill should have different locations and techniques from the previous drill. Do not use the same fire hose stations for each drill.

3-39. **Required Features.** Effective fire drills do not happen automatically. Careless effort will result in useless drills which do not improve the crew's capability, or even bad drills which train poor habits. Each fire drill should include training elements which touch on all phases of fire fighting. One example is training in combating a deep fat fryer fire. These drills will ensure personnel know how and when to secure the fryer and extinguish the fire. As a fire fighting party improves, realism can be incorporated. One realistic means of insuring the crew knows initial reporting procedures is to use red rags and signs to indicate presence and class of a fire. When a crewmember discovers the fire they must report the simulated fire using proper procedures.

- Time compression is the most important feature to incorporate. A fire can grow from a tiny flicker to a life threatening blaze in a few minutes. Every delay in detection, notification, fire fighting, and space isolation could cost a life or another burned out compartment. Drills must be practiced at real time speed. This creates two important conditions: the urgency of the situation and the inevitable problems with donning personal protection.
- Training in the use of the EEBD should be emphasized. Quickly donning the EEBD should be stressed as a way of saving time and improving the chances of survival when escaping a fire or smoke-filled space.
- The effects of smoke must also be included. These effects include the loss of visibility, loss of staging areas, loss of equipment in lockers which cannot be reached, and the extra confusion caused by all the above.
- Cascading casualties are also common in fires, as a fire spreads or damages vital services. Realistic, effective drills shall include these effects.
- Machinery space fires can grow out of control in seconds. For this reason, abandon the space evacuation drills should be conducted. Such drills should be focused on how to abandon the spaces quickly and safely.

3-40. **Critique.** Critiques after every fire drill will help ensure that the maximum learning takes place. They should examine the underlying causes for successful or failed drills. They should include a thorough discussion of the rationale for each decision made on attack points, ventilation, and so on. This is the perfect time to review the results of fire fighting equipment performance/accountability and plan any repair/procurement.

FIRE FIGHTING PROCEDURES

3-41. The designated Fast Reaction Team will proceed directly to the scene of the fire. The Fast Reaction Team will consist of the On-scene leader and investigators.

3-42. The On-scene leader is in charge of the scene and must quickly assess the situation and decide on an immediate course of action. Containment of the fire is of primary concern initially. Boundaries must be quickly established to contain and halt the spread of the fire.

3-43. Investigators will also quickly assess the situation and establish boundaries as they investigate the areas surrounding the fire. Investigators will search all areas surrounding the space of the fire and report their findings to the On-scene leader.

3-44. Size up the fire – The on-scene leader should determine, as quickly as possible;
- The class of fire.
- The appropriate extinguishing agent.
- The appropriate method of attack.
- How to prevent extension of the fire.
- The required manpower and firefighting assignments.

Communications

3-45. Communications between the master and firefighting teams should be rapidly established. Communication is vital to an effective battle plan to combat the fire. Effective communication allows the Master to utilize all the resources of the vessel in cohesion and greatly increases the effectiveness of the crew. Hand-held two way radios are effective for establishing communications, however a secondary method of communications should be planned for and utilized during drills to prepare for the possible loss of radio communications. A messenger is in many instances the best for this purpose.

Staging Area

3-46. The On-scene leader will determine a staging area location in a smoke free area, as near as possible to the fire area. Ideally it will be an open deck location, windward of the fire. All supplies and back ups needed to support the firefighting effort should be brought to the staging area. The fire teams will be directed to the staging area after donning the FFE. Upon their arrival the On-scene leader will commence a plan to combat the fire.

3-47. The attack on the fire should be started as soon as possible.
- **Direct Attack.** Firefighters advance to immediate area and apply the extinguishing agent directly into the seat of the fire.
- **Indirect Attack.** This method is employed when it is impossible for firefighters to reach the seat of the fire. Must be made air tight as possible.

3-48. **Ventilation.** The actions taken to release combustion gases trapped within the ship and vent them to the outside atmosphere. Most fire fatalities result from asphyxiation by combustion gases or lack of oxygen, and not by burning. Ventilation is used only when a direct attack is made on the fire.

3-49. Rescue of trapped personnel is an extremely important aspect of every firefighting operation. Rescue may be the first step in the operation, or it may be delayed because of adverse circumstances.

FIGHTING CLASS B FIRES IN ARMY VESSELS

NOTE: There can be no substitute for prudent, common sense, on-the-scene decisions which may dictate variations to this guidance. Restricted maneuverability may also require departure from this guidance. While the personnel responsibilities and scenarios described in this manual may be developed for use in individual fire doctrines on most ships, they may not be applicable in every situation. The ship's associated equipment isolation and controls lists are available in Damage Control.

3-50. This section covers Machinery Space Fire Fighting Doctrine for Class B Fires in Army Vessels. Topics covered include:
- Fire prevention
- Fire fighting systems capabilities and limitations
- Considerations necessary in choosing the correct fire fighting equipment
- Actions necessary both internal and external to the affected space in the case of a major oil leak, a class B fire, and a fire which grows out of control

PREVENTION

3-51. The following eight principles shall be enforced to reduce fire hazards.

- **Principle 1:** Inspect equipment regularly and report conditions to the vessel master.
- **Principle 2:** Properly stow and protect all combustibles.
- **Principle 3:** Test and inspect flammable systems after repairs.
- **Principle 4:** Conduct fire drills on a regular basis.
- **Principle 5:** Educate all personnel in the reduction of fire hazards and perform frequent inspections. Enforce fire prevention policies and practices.
  - Maintain proper covers on flammable liquid strainers and keep sounding tube caps in place and isolation valves closed. Ensure all flammable liquid sounding tubes terminating in machinery spaces are properly equipped with ball check valves, isolation valves, and sounding tube caps.
  - Immediately stop oil leaks and repair.
  - Wipe up spilled or leaked oil.
  - Keep ventilation ducts free of oil residue.
  - Keep bilge's free of oil and trash.
  - Prevent stockpiling of excess or unauthorized flammables.
  - Ensure uptake spaces are not used as storerooms for combustible materials.
- **Principle 6:** Properly maintain all firefighting equipment.
- **Principle 7:** Operate and maintain systems and equipment according to authorized procedures.
- **Principle 8:** Properly maintain all machinery space damage control closures and fittings.

CLASS B FIRE SCENARIO

3-52. A class B fire can result from any pooled oil and can quickly develop from an oil spray or atomized fuel. The following guidelines are provided for a class B fire.

- **Report the Fire.** The fire should be reported immediately to the Space Supervisor to allow for concurrent actions. When fire or smoke is reported, and as soon as fire fighting and plant securing efforts allow, personnel in the space should obtain and carry an EEBD. EEBDs are designed for escape only and shall not be used for fire fighting purposes.
- **Man Foam Proportioning Station.** Man the machinery space fire stations upon notification of a fire in the space.
- **Size Up the Fire.** Assess the size and location of the fire. If the fire is localized, activate the AFFF hose reel and extinguisher (where installed) and attack the fire by advancing toward it and extinguish the bilge fire at the deck or bilge areas. One AFFF hose reel and portable extinguisher require two persons. When within 20 feet of the oil spray or leak, direct as needed to prevent re-ignition. Discharge AFFF to the residual fire on the deck, in the bilge, and over the surrounding area until the fire is extinguished.
- **Secure the Oil Source.** The leak should be stopped as quickly as possible. This can be done locally or remotely by closing system cutout valves or shutdown valves, therefore isolating the leak. Even the application of a rag or bucket can significantly reduce the flow of oil and deflect it away from the hot surface.
- **Concurrent Actions.** As time and personnel permit, the following concurrent actions should be accomplished.
  - Deploy additional portable extinguishers to the scene of the fire.
  - **Secure the plant and operating machinery in the affected space.** Start or maintain equipment in unaffected spaces to maintain propulsion, electrical power, and fire main.
  - Set ventilation according to the following:
- **In affected machinery space:** Set negative ventilation (exhaust on high and supply on low). On ships with interlocked fans and remote controls with emergency exhaust button. Set emergency exhaust on high and supply off. On ships with fans interlocked through a local master switch but

with independent control on controllers inside the space, set negative ventilation. On other ships with interlocked fans, the ventilation system shall remain operating.

- **In unaffected machinery space:** Set positive ventilation (supply on high and exhaust off). Setting positive ventilation is intended to prevent smoke on the damage control deck from entering unaffected spaces. On ships with fans interlocked through a local master switch inside the space but with independent control on controllers, set positive ventilation. On other ships with interlocked fans, the ventilation system shall remain operating. If smoke is ingested into adjacent machinery spaces from the weather, shift ventilation supply to intakes on opposite side of ship, if installed, or maneuver the ship to clear the vent intakes or secure ventilation. Watch standers may require BAs to prevent premature evacuation of unaffected spaces because of smoke. Isolate the affected space with the exception of fire fighting equipment, lighting, and ventilation. Set fire and smoke boundaries around the affected space to prevent the spread of fire and smoke throughout the ship. The ship may want to set general quarters to facilitate the complete isolation of the affected space and the rapid establishment of fire and smoke boundaries. In setting boundaries, consideration should be given to the trade-off between impeding personnel egress versus spread of smoke to unaffected spaces.

OUT-OF-CONTROL CLASS B FIRE SCENARIO

3-53. A class B fire, especially one that has burned for a period of time or is fed by an un-securable oil source, can become out of control within seconds. When this happens, operating machinery and the plant should be secured and the space evacuated. The following guidelines are also provided for consideration when faced with an out-of-control fire.

3-54. **Size of the Fire.** If the fire occupies a large area, is fed by an oil source which cannot be secured, or is threatening fire fighting and escape, the space should be evacuated. Even a small fire, if not extinguished rapidly, can generate large volumes of smoke and toxic gases that can force a space to be evacuated.

3-55. **Evacuation.** Once the decision is made by the officer in charge (OIC) to evacuate the space, the fixed fire fighting system will be activated, if installed, and all personnel should don their EEBD and exit the nearest safe access. To prevent running the system dry, operate the system no longer than 4 minutes. Never operate the system when the concentrate level in the tank sight glass is not visible. Immediate manning of the AFFF proportional is essential to expedite tank replenishment. Access doors, hatches, and scuttles shall be secured when all personnel are out of the space. At this time, ventilation shall be secured for ships without the fixed firefighting system. The escapees should congregate at a safe, predetermined location outside the space, where EEBDs can be removed and a muster taken. A safe location is outside fire and smoke boundaries or a weather deck. Notify the OIC that the following actions were taken.

- Lighting to space has remained on.
- Ventilation to the space has been shut down.
- Fixed firefighting system has been activated, if installed.
- The space is evacuated and all personnel are accounted for.
- The Space Supervisor has completed briefing the OIC on the location of the fire and plant status.

FIRE FIGHTING SYSTEMS CAPABILITIES AND LIMITATIONS

3-56. All ships are provided with one or more of the fire fighting systems or equipment as described in the following paragraphs. Each has capabilities and limitations which shall be known and understood by fire fighting personnel to ensure quick and proper selection of equipment. Fixed Gaseous Fire Fighting Systems are the primary fire fighting agent for extinguishing class B fires in machinery spaces except for those cases where the fire can be extinguished by handheld equipment.

FIRE PUMPS (FIRE MAIN SYSTEM)

3-57. Centrifugal pumps are installed for seawater supply to the Fire main system. The fire main system will be kept intact so that water is available for cooling and the production of AFFF. Water is useful for cooling hot bulkheads in those spaces adjacent to the fire and extinguishing ordinary combustible (class A) fires. When a hose line attack is needed to extinguish a flammable liquid fire and AFFF is not available, high velocity water fog may be used. However, time to fight the fire will be longer, more firefighters will be needed, increased fire damage can be expected, and a greater risk of re-flash will be present.

AQUEOUS FILM FORMING FOAM (AFFF)

3-58. AFFF concentrate, when mixed with water, creates foam for application to surfaces. AFFF has a shelf life of 25 years when the storage container is not opened and exposed to contamination. It can be applied by a separate fire plug and hose with portable foam proportioned (inline eductor) It is effective on bilge fires to smother burning liquids, prevent large scale re-flash and for use during space reentry. When a hose line attack is made to extinguish a flammable liquid fire, AFFF shall be used unless expended or out of commission. In this event, use of water fog is acceptable.

REQUIRED CAPABILITIES

3-59. The following is required of all systems:

- All systems are capable of total fire extinguishment and provide 15 minutes of re-flash protection.
- The systems will be designed to be acceptable for marine use, compatible with existing ships systems and operational environments. The normal operating temperature range of the spaces protected will vary from 32 to 130 degrees F.
- All systems will be equipped with time delays and pre-discharged alarms based on personnel evacuation time or to prepare the hazard area for discharge. Also, time delays will have the capability to be manually by passed (exception: time delays are not required for flammable liquid storage).
- All systems will be manually activated. Actuation will be accomplished by either mechanical or pneumatic means. Electrical actuation will not be allowed. Also, for those cylinders located within the space protected, automatic activation by a heat actuator will be provided.
- Natural and forced ventilation to the protected spaces will be secured prior to system activation. Automatic shutdown of powered ventilation and procedures for securing of natural ventilation prior to system activation will be required.
- All internal combustion engines located in protected spaces, which draw intake air from within the protected spaces, will be equipped with shutdown devices which are automatically actuated in the event of the fire extinguishing system activation. All internal combustion engines located in protected spaces, which draw intake air from outside the protected spaces, will be equipped with shutdown devices which are manually actuated in the event of the fire extinguishing system activation.
- All agent storage cylinders will be securely supported and rigidly fastened and be equipped with pressure gauges and magnetic liquid level indicators.
- System designs will be based on watercraft being fully operational, as opposed to cold iron. Maintenance and training requirements will include winterization, lay-up, and re-activation procedures.
- System installation will permit normal operations required in walking and working areas without undue interferences, clear headroom of 6 feet 3 inches is required in all walking and working spaces.
- System details. Design concentration, total flooding quantity, and discharge rate for fire extinguishing agent will be based on the following:

3-60. The minimum amount of agent concentration supplied by each system for fire extinguishment will be based on cup burner extinguishment concentration plus a 40 percent safety factor at 20 degrees C (68 degrees F). - In no case will the design concentration exceed the lowest observed adverse effect level (LOAEL) of 10.5 percent at the highest ambient temperature expected in the hazard areas at 54 degrees C (130 degrees F).

3-61. The total flooding quantity required will be calculated in accordance with NFPA 2001, Section 3-5.

- The system will discharge within 10 seconds.

3-62. The system will incorporate the following features specific to Army watercraft.

ALARMS

Audible

3-63. All protected spaces will be equipped with pre-discharge alarms (audible and visual). The alarms will be conspicuously marked. The audible alarm will sound for the required duration prior to release of the agent into the space. The audible alarm will be sirens powered only by the agent released. Also install an electric alarm bell outside each protected space. The bell will be activated by means of a pressure switch actuated by the release of agent. This alarm bell will sound continuously until manually reset. The alarm bells will be powered by the watercraft's emergency power source. This is to warn the ship's personnel that the system has been deployed. For engine room spaces, the sirens must be audible above the sounds of operating machinery and be audible in the control room (where applicable), as well as in the machinery space.

Visual

3-64. All protected spaces will be equipped with visual alarms. The visual alarms will be an amber strobe light activated by means of a pressure switch actuated by the release of agent. Multiple spaces protected within a compartment will require a strobe light for each space. Lights will be powered by the watercraft's emergency power source. Paint lockers are not required to have visual alarms.

Time Delays

3-65. All systems will be fitted with an approved time delay so that the alarms will operate before the agent is released into the space. Also, time delays will have the capability to be manually by-passed. Paint lockers are not required to have time delays. Time delay bottles contain CO_2 gas when released, the amount of CO_2 that is released is sufficient enough to cause unconsciousness and death to all individuals within the space.

CONTROLS AND VALVES

3-66. Where necessary, excessively tight compartments such as small paint lockers will be provided with suitable means for relieving excessive pressure accumulating within the compartment when the agent is released and allowing for proper agent progressive mixing within the protected space atmosphere. Controls and valves for operation of the system will not be located in any space that might be cut off due to a fire in any space protected (exception: paint lockers). Some suitable means are as follows:
- Manual and automatic actuation control stations
- Remote and automatic shutdown devices
- Audible and visual activation alarms

3-67. Two independent manual actuation control stations are to be provided, one of them being positioned at the cylinder storage location and the other in a readily accessible position as convenient as practical to the main escape from the space. Also, for standardization, provide a third remote manual actuation station outside on the main deck for the main propulsion engine rooms, as presently located on the majority of watercraft. These actuations control stations will be mounted in one corrosion-resistant watertight enclosure, capable of withstanding heavy sea conditions and will be quick-acting to open. Placard directions will be mounted on the inside cover.

3-68. Systems will be actuated from the actuation stations by two control levers. One control lever operates the stop valve to the space and the other control lever is a separate control that releases the agent. These controls will be in individual pull boxes clearly identified for the particular space. Actuation stations will be conspicuously mounted to facilitate operation in an emergency. Actuation stations will be standardized (same manufacturers) throughout the fleet (exception: paint lockers will be actuated by one control level releasing the agent (break glass, pull lever).

3-69. All systems will be manually controlled. For systems where the cylinders are stored within the protected space, the system will be fitted with an automatic heat actuator device. This device, in the event of an undetected fire in the protected space, will allow the system to safely release the contents of the cylinders into the protected space. These spaces will also have two independent manual-operated releases as previously specified.

MARKINGS FOR FIRE AND EMERGENCY EQUIPMENT

3-70. All system warning placards will be of the phenolic (plastic molded) type. Markings will be permanently attached, displaying 1/2 inch red lettering on a white background. System piping will be marked in accordance with TB 43-0144. The distribution line valves of all extinguishing systems will be plainly and permanently marked indicating the spaces served. Complete and simple instructions for operation of system will be posted in a conspicuous place at or near the pull boxes, stop valve controls, and in the cylinder storage rooms. Systems where cylinders are stored outside of the protected space will include a schematic of the system and instructions for alternate method of discharge should the manual release or stop valve fail. The discharge and pre-discharge alarms will be conspicuously identified in red letters at least 2 inches high.

CHOOSING CORRECT FIRE FIGHTING EQUIPMENT

3-71. The proper choice of fire fighting equipment should be based on an on-the-scene estimate of the situation. This estimate should be done quickly and should consider the volume of flammable liquid released and its form (atomized or spilled); the area occupied by the flammable liquid (confined or unconfined); the ability to quickly secure the oil source; and how rapidly flame, heat, and smoke are threatening fire fighting and escape. The following general guidelines are provided for consideration when selecting the proper class B fire fighting equipment.

- **Small Pool Fires (Less Than 10 Square Feet).** Use readily accessible portable extinguishers or AFFF hose reel. CO2 portable extinguishers will not be used on fires greater than 4 square feet.
- **Oil Spray Fires.** An oil spray fire resulting from the ignition of atomized flammable liquids should not be attacked. Loss experience and fire testing have demonstrated that a pressurized release of a flammable liquid can create a fire that is unapproachable. Life threatening conditions created by extreme heat, smoke, and toxic gases can occur, especially on the upper level, in as little as 60 seconds. Under such conditions, the only prudent action, time permitting, is to secure the propulsion plant, don EEBD, and evacuate. Oil spray fires may occur around fuel and lube oil strainers, recently repaired flanges and valves, and flexible line failures. An oil spray fire can grow out of control within seconds. Such fires are commonly fueled by an oil source which cannot be quickly and completely secured, including those fires with a fuel source from an oil tank sounding tube terminating in a machinery space, and will most likely grow out of control thereby requiring space evacuation. Fires which spread to overhead insulation and cables, or which produce sufficient products of combustion (flame, heat, smoke, and gases) can also force space evacuation.

SMOKE CONTROL

3-72. Smoke control is comprised of the following areas.

- **Ventilation.** The operation of ventilation systems is described where required in this manual. Each ship shall supplement its SOP with a list of fans and their controls to be secured for a designated fire and buffer zone. Weather deck supply intake and exhaust discharge locations shall be listed. The location of controllers, their designation, and area served shall also be listed.
- **Smoke Boundaries.** The use of smoke boundaries around the affected space can effectively limit the spread of smoke and provide controlled areas for the staging of fire fighting personnel. They shall be set quickly using, as a minimum, fume tight boundaries which each ship shall clearly identify. The objective of primary smoke boundaries is to first establish a buffer zone by closing those hatches and doors immediately adjacent to the access for the affected space. Smoke curtains may be used where hatches and doors may be required to remain open for fire fighting purposes. This buffer zone shall be a dead-air area. Only personnel with Breathing Apparatus (BA) are allowed to enter this area once it is established. BAs should be used when smoke is present or when ordered by the scene leader. A second boundary shall be set around the buffer zone to check the spread of smoke and provide a safe area for fire fighting personnel without BAs. Each ship shall supplement its SOP with a list of designated smoke boundaries for machinery spaces.
- **Maintaining Fire and Smoke Boundaries.** Once a machinery space has been evacuated, fire and smoke boundaries should be maintained. At the time of reentry, firefighters may encounter a back draft explosion as accesses to the affected space are opened and hot fire gases are relieved onto the

damage control deck. Firefighters should use caution to position themselves to the side of the access when the door, hatch, or scuttle is initially opened.

SPACE ISOLATION

3-73. The complete isolation of the affected space, with the exception of lighting, is necessary to prevent a fire from intensifying due to the addition of flammable liquids and oxygen, and to reduce the electrical hazards. Each ship shall supplement its Standard Operating Procedures for shipboard emergencies with a list of local and remote controls (valves, switchboards, circuit breakers, and so forth), for rapid space isolation. The designation, location, function, and area served by each control shall be provided.

3-74. The following areas shall be considered when isolating the space:
- **Mechanical.** Every effort shall be made to secure and isolate those systems, machinery, and tanks that have the potential to feed or otherwise contribute to the intensity of the fire. These include, in priority, those systems where action should first be taken.
- Fuel transfer, service and stripping pumps, and centrifugal purifiers.
- Fuel systems, storage, and service tanks.
- Lube oil pumps and centrifuge purifiers.
- Hydraulic systems.
- Lube oil tanks.
- Air compressors.
- **Electrical.** To the extent possible, all electrical equipment, with the exception of lighting, shall be secured from outside the affected space at the ship's service, and emergency switchboards, load center, or distribution panel. The switches, circuit breakers, and fuses necessary to do this shall be clearly identified.
- **Fire Boundaries.** Fire Boundaries shall be established around the affected space to contain the fire and ensure designation of adjacent spaces to be observed for hot bulkheads. These boundaries are generally the watertight bulkheads and decks immediately adjacent to the affected space. The minimum degree of tightness for a fire boundary is fume tight. The ship may set general quarters to rapidly establish fire boundaries. Each ship shall supplement its Standard Operating Procedures for shipboard emergencies with a list of designated fire boundaries for machinery spaces.
- **Fuel Tanks.** Transfer of fuel to a safe location to remove fuel tank contents puts the empty fuel tank at maximum risk to fire. Therefore, transfer of fuel from the fire area should not be attempted. In summary, the only action necessary to prevent tank contents from contributing to a machinery space fire is to isolate and secure the fuel system.

3-75. When initiating action to secure and isolate the foregoing, the following factors shall be considered.
- Not all of the above have remote securing or isolation capability. As such, much local securing or isolating shall be accomplished as soon as possible together with the start of fire fighting actions. As a minimum, local securing of systems shall include tank and bulkhead boundaries. Familiarity with location and type of local securing and isolating capabilities, and casualty control procedures such as those contained in the applicable propulsion plant manuals is required.
- Where remote capability is provided for any of the above, it is most likely located within or immediately adjacent to the access to EOS and at the access to the machinery space on the damage control deck. Ensure system isolation by visual or operational verification of all remote actuators.
- Care shall be exercised to prevent cascading casualties to equipment in unaffected spaces necessary to maintain propulsion, electrical power, and fire main pressure. Air systems, air compressors, and fuel tanks located close to space boundaries are of particular concern. Communication with other machinery spaces is essential to reduce the potential for casualties.

REENTRY

3-76. Reentry to a machinery space that has been evacuated because of fire is the most critical part of the fire fighting evolution and potentially the most dangerous. The primary function of the reentry team is to attack and extinguish the fire, ensure the source of oil is secured, and cool the space so ventilation may be started.

GENERAL GUIDELINES FOR REENTRY

3-77. When the fixed firefighting system has been activated:

- If the evidence is that the fire is extinguished, do not attempt reentry for at least 15 minutes after the discharge.
- If conditions in the affected space indicate that the fire has not been extinguished and continues to grow after the fixed firefighting system has been discharged:
- Feel bulkheads for temperature near the desired access.
- Monitor exhaust vent discharge for smoke.
- If CO2 equipped vessel, ventilate space and test BEFORE entry attempt.
- If FM-200 equipped vessel, test space using Damage Control hand pump with ampulets.
- Monitor conditions through the Engine Operating Station (EOS) windows or peephole in escape doors.
- After reentry to the space, fires shall be extinguished. Reopen when the fire is out, re-flash watch is set, and fire overhauled. Ensure all sources of fuel are secured and covered with AFFF. To conserve AFFF, saltwater hoses should be used to cool the space after the fire is out. It should be assumed that AFFF hose reels in the space have been damaged by the fire and they should not be relied on until it can be established that the system has not been damaged by fire.

3-78. When FM 200 is not installed:

- Electrical isolation, with the exception of lighting, shall be completed as specified in EOCC immediate actions. Electrical isolation, although ongoing, should not delay space reentry.
- Reentry should be made through the access, main door, hatch, or escape trunk, whichever is not obstructed by the fire. The conditions in the affected space should be checked before entry by feeling bulkheads for temperature near the desired access, monitoring exhaust vent discharge for smoke, and monitoring operating conditions through the EOS windows or peephole in escape trunk doors.
- Repeated efforts may be necessary to gain access to the space. The nozzle man uses the reentry hose wide-angle fog to cool metal surfaces and protect himself. It should be assumed that the hose reels in the space have been damaged by the fire and they should not be relied on for use until it can be established that the system has not been damaged by fire.
- Once inside the space; locate, extinguish, and report: fire out and set re-flash watch. Report that re-flash watch is set. Overhaul the fire. Secure and cover all flammable liquids with AFFF. Allow the space to cool. To conserve AFFF, saltwater hoses should be used to cool the space after the fire is out.

DESMOKING, ATMOSPHERIC TESTING, DEWATERING, AND REMANNING

3-79. After the fire is out, the space should be made safe and ready for re-manning. A re-flash watch shall be posted with portable AFFF to quickly extinguish any fire which may reignite.

DESMOKING

3-80. When a flammable liquid fire has been extinguished, combustible gases may be present. Operating electric controllers to start fans may ignite these gases. De-smoking with the installed ventilation system can proceed with minimal risk when the source of fuel is secured, the space allowed to cool, all fuel washed into the bilges, and no damage sustained to the electrical distribution system. Another method is to conduct de-smoking with fire hoses. Clearing the space of smoke should commence as soon as the space has cooled sufficiently so there is no danger from re-ignition. Circuit breakers and other protective devices which tripped automatically shall be left in the tripped position until system damage has been assessed. Examine the electrical distribution system and if possible reestablish power to the installed ventilation fans. If fully operational, run all fans on high speed for a minimum of 15 minutes to remove smoke and toxic gases. If the installed system is partially or fully inoperable, de-smoking will take longer, but can be accomplished by using portable blowers, operable installed fans, or positive pressure from adjacent spaces and opening access to the affected space. The safest way to de-smoke machinery spaces on ships without the fixed firefighting system is to exhaust with portable blowers or to use positive pressure from adjacent spaces. These methods reduce the risk associated with igniting flammable liquids that have not been vapor secured with AFFF.

ATMOSPHERIC TESTING

3-81. De-smoking shall precede atmospheric testing because combustible gas analyzers will not operate reliably if the sensor is exposed to excessive moisture or comes in contact with particulate found in post-fire atmosphere. When the space is clear of smoke, test for oxygen, combustible, and toxic gases. The level of oxygen shall be between 19.5 and 23.5 percent, combustible gases shall be less than 10 percent of the lower explosive limit, and all toxic gases below their PEL values before the space is certified safe for personnel without BAs. Shipboard personnel authorized to conduct post fire atmospheric tests for the purpose of certifying the space safe for personnel are Gas Free Engineering personnel. The repair party post fire gas free test assistant is not authorized to make safe for hot work gas free tests unless the assistant is qualified. If FM-200 has been discharged, a test for hydrogen fluoride shall also be conducted. Tests should be conducted near the center and all four corners on each level. At least one satisfactory test shall be obtained at each location tested. Instruments used shall be approved by the National Institute of Occupational Safety and Health, Mine Safety Health Administration.

REMANNING

3-82. Once the space is certified as safe, re-manning can begin. Operation of equipment and desolation of mechanical and electrical systems shall be considered only after a careful assessment of damage.

PERSONNEL PROTECTION AND FIRE FIGHTING EQUIPMENT

3-83. The proper use of personnel protection and fire fighting equipment is required to reduce the risk of injury and facilitate extinguishing the fire. Some general considerations for those individuals who enter the space are the following.

> **NOTE**
> Oxygen Breathing Apparatus (OBA) / Self-Contained Breathing Apparatus (SCBA) will be referred to as BA in this chapter.

- **Breathing Apparatus (BA).** Breathing Apparatus (OBA / SCBA), with voice amplifier if on ship's allowance list, should be worn by all personnel within buffer zone or when entering the affected space until the atmosphere is declared safe. When smoke is present, Breathing Apparatus activation should be ordered and reported to the OIC.
- **Clothing.** Fire fighters ensembles shall be appropriately pre-positioned to be readily available to fire party personnel when the fire party is called away. Personnel required to wear the firefighter ensemble are the scene leader, nozzle men, and hose men. Support personnel such as phone talkers, plug men, electricians, and medical personnel, outside the fire boundary, shall wear battle dress uniforms, anti-flash hoods, and gloves.
- **Hoses.** As a minimum, a single attack 1 1/2-inch saltwater hose shall be used by the reentry team. The hose and nozzle provide added protection for the nozzle-man and hose tenders. Before the single attack hose enters the space, a second backup attack saltwater hose should be manned and charged to render assistance. When assigned by the scene leader, each hose team will be led by an attack team leader. Sufficient distance shall be maintained between the first and second hoses to prevent maneuverability and fire fighting progress from being impaired. Inasmuch as reentering the space may be a lengthy and awkward process, saltwater hoses should therefore be used to cool access doors, hatches, and scuttles. Saltwater hoses should not enter the space where it will impair the effectiveness of the AFFF hose teams. To conserve AFFF, hoses equipped with inline eductors can discharge saltwater if pickup tubes are removed from AFFF 5-gallon cans. The eductor will continue to function with reinsertion of the pickup tube into the AFFF containers.

BREATHING APPARATUS (BA)

3-84. Army vessels have currently two distinct types of breathing apparatus, the Self-Contained Breathing Apparatus (SCBA) and the oxygen generated breathing apparatus (OBA). The SCBA will replace the OBA in the future. Information for both types is present for operational and training information.

SELF-CONTAINED BREATHING APPARATUS (SCBA)

3-85. The SCBA provides the firefighter with air for breathing while in a fire or another hazardous environment. It consists of a composite or metal cylinder for high pressure (4,500 PSI) air storage and regulators to reduce air pressure to useable levels. Air is provided to the user at above ambient pressure to provide a positive pressure in the face piece and prevent hazardous gases from entering. The SCBA can be recharged using refill connection. The quick-fill consists of a quick disconnect. Specifications may be found in Appendix C.

3-86. To determine which size is best for you do the following:

- Starting with a large face piece, extend the straps, place your chin in the chin cup and pull the straps over your head.
- Tighten the top straps first and then the lower or chin straps. Ensure that the "figure 8" in the middle of the straps is centered on the crown of your head.
- Place the palm of your hand over the opening in the front of the face piece where the regulator installs. Inhale and listen/feel for leaks. The mask should collapse against your face. If the mask leaks, try tightening the straps.
- If this does not help or causes discomfort then you may need to try the small mask. If the mask feels too small on your face then you might try the extra large face piece.
- Repeat the leak check in each case.

SCBA DONNING PROCEDURE

3-87. Inspect SCBA. Perform a quick visual inspection for any obvious damage that would preclude safe and proper use of the SCBA. Check the following parts for damage:

- **Face piece:** rubber deterioration, cracks, tears, holes, tackiness
- **Head Harness:** breaks; missing straps; loss of elasticity; buckles deformed, corroded, damaged or missing; strap serrations worn
- **Lens:** cracks, scratches, loss of tightness to face piece, regulator inlet coupling deformed or damaged
- **Backpack:** cuts, tears, abrasions, signs of chemical or heat damage, inoperative buckles, damage to wire frame
- **Cylinder:** minimum pressure allowed is 4000 psi; check cylinder for charring, dents, gouges or cuts that may have penetrated fiberglass or carbon fiber
- **Hoses:** cuts, cracks, abrasions, bulges, wrinkles, loose or inoperative connections
- **EZ-Flow Regulator:** Check casing for damage, dirt or debris. Actuate purge valve, air saver switch and removal lever. Check sealing gasket.

3-88. Don SCBA (coat method or over-the head)

- Most of the weight (85%) should be carried on the waist/hips
- Check all straps for correct adjustment

3-89. Don Face Piece

- Spread face piece straps from inside with thumbs
- Place chin in chin cup
- Pull harness over head and smooth straps
- Ensure all hair is away from seal
- Tighten neck straps first, then temple straps

3-90. Perform Negative Pressure Check

- Place hand over face piece opening for second stage regulator
- Inhale and hold your breath
- Listen/feel for inward air leakage
- Adjust face piece as needed until seal is maintained

3-91. Open Air Cylinder Valve

- First depress air saver switch
- Ensure purge valve is closed,
- Open cylinder valve, close 1/4 turn, and listen for Vibralert to sound during initial equalization *(the bell alarm may or may not sound at this time)*

3-92. To Begin Operation ("go on air")

- Attach Second-Stage regulator to face piece lock in place
- Inhale sharply to begin flow of air
- Breathe with purge valve open to experience breathing in free flowing air
- Close the purge valve, press air saver switch and disconnect regulator from face piece
- Wait to engage the air saver switch until just prior to pulling the regulator away from the mask

REFER TO SPECIFIC MANUAL INSTRUCTIONS FOR DETAILED GUIDELINES.

OXYGEN BREATHING APPARATUS (OBA)

3-93. A thorough knowledge of the OBA (figure 3-2) and its function is vital to ensure the safety of the operator until all vessels are issued the replacement SCBA. This section provides a brief description of the functional relationship that exists between the major parts and assemblies that comprise the BA. It also discusses their interactions as they relate to forming the closed breathing system.

3-94. The OBA consists of the following nine operational parts and assemblies.

- Face piece
- Inhalation valve
- Inhalation tube
- Breathing bag
- Pressure relief valve
- Canister
- Exhalation tube
- Exhalation valve
- Combination valve assembly (The combination valve assembly is composed of the inhalation valve, the exhalation valve, a speaking diaphragm, and the valve body or housing)

DESCRIPTION

3-95. The following describes the operation of the OBA (figure 3-2). The key elements of the system are described below.

- Air Flow. The face piece (1), worn on the head, covers the face of the user and seals it from the external atmosphere. A speaking diaphragm is built into the face piece to allow conversation while the apparatus is in use. As the user exhales, moist breath is carried down the exhalation tube (7) and into the bottom of the canister (6). It then rises upward through the chemicals in the canister, producing oxygen. This oxygen passes into the breathing bag (4) and then up through the inhalation tube (3), where it is drawn into the face piece. The inhalation and the exhalation check valves (2 and 8), located in the combination valve assembly (9), permit flow in only one direction. The pressure relief valve (5) in the breathing bag automatically relieves excess pressure.
- Face piece. The face piece is made of rubber in a face seal configuration. A wide angle lens made of scratch resistant plastic provides the wearer with good visibility and reduces any claustrophobic effects. The face piece also contains a rubber nose cup. The combination valve assembly and speaking diaphragm are in a hard plastic housing at the bottom of the face piece. The combination

valve assembly contains the inhalation and exhalation check valves. Oxygen entering the face piece is directed over the face piece lens to reduce fogging. Oxygen then enters the nose cup through two check valves in the nose cup. Exhaled breath leaves the nose cup through the exhalation check valve.

● Timing Device. To warn the user when the oxygen supply is running low, a timing device is attached to the OBA. The timing device consists of a timer and bell. The time is graduated in minutes and may be set for any fraction of 60 minutes. During use the time is set for 30 minutes. After the set time has expired, the bell will ring for a minimum of 10 seconds to warn the wearer to return to fresh air.

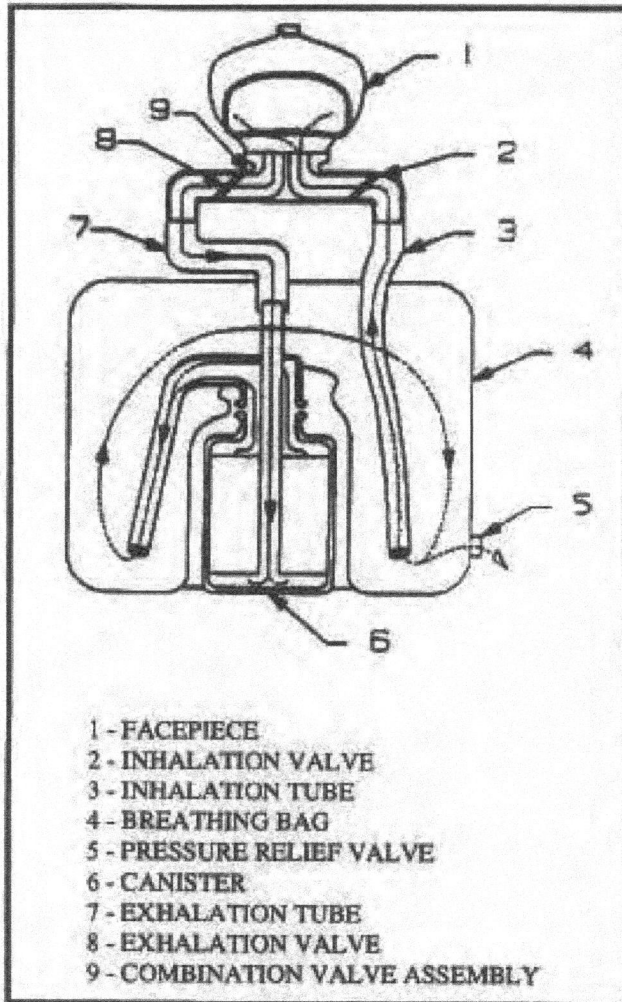

1 - FACEPIECE
2 - INHALATION VALVE
3 - INHALATION TUBE
4 - BREATHING BAG
5 - PRESSURE RELIEF VALVE
6 - CANISTER
7 - EXHALATION TUBE
8 - EXHALATION VALVE
9 - COMBINATION VALVE ASSEMBLY

Figure 3-2. OBA operational parts and assemblies

OBA TRAINING CANISTER KIT

3-96. **Training Canister.** Training canister kits are used to instruct personnel on the use and operation of the OBA. The operation of the fully assembled training canister is identical to that of the quick start canister.

However, the chemical in the training canister is not the same as the chemical used in the quick start canister. The chemical in the training canister absorbs CO2, but it does not produce oxygen. The amount of chemical in the training canister is enough to absorb CO2 for forty trainees not doing firefighting work.

3-97. **Firing Mechanism and Candle.** An oxygen-generating candle provides the oxygen for an OBA fitted with a training canister. The replaceable candle generates a five minute supply of oxygen (10 liters), beginning within 15 seconds after the candle has been fired. The firing mechanism is located on the bottom of the canister directly beneath the candle. The firing mechanism is removable. Pulling the lanyard causes a spring loaded hammer to strike a primer and ignite the candle.

SAFETY PRECAUTIONS AND GENERAL INFORMATION

SAFETY PRECAUTIONS

3-98. Follow all safety precautions to ensure the safety of the wearer while using the Type A4 OBA (figure 3-3). The manual start method is the Army's only recommended way to start the OBA Canister. If the canister is fired while the foil seal is still in place, pressure will build in the canister, causing the foil to rupture.

> **Warning**
> **Quick Start Candle method is unsafe for use.**

EQUIPMENT DESCRIPTION

3-99. The BA is a personnel protection device which, when placed into operation, provides the wearer with an isolated, self generating oxygen supply for 45 minutes. It is worn when the atmosphere in a space is dangerous to personnel. The 45-minute time period provided by the OBA consists of a 30-minute arrival and on-station interval along with a 15-minute supply of reserve air allotted for leaving the scene. The OBA, wearer, and canister form a closed system during operation. Chemicals in the canister remove CO2 and moisture from the user's exhausted breath and replenish it with O2. This allows the wearer to survive and work in a toxic atmosphere. The OBA will not protect the wearer from any skin absorbed hazards.

ACCESSORIES

3-100. **Wrench.** The wrench is used to remove the valve cap from the speaking diaphragm on the face piece assembly.

3-101. **Spectacle Kit.** (Figure 3-5) For personnel who wear eyeglasses, a spectacle kit is available for installation of prescription lenses in the face piece. The spectacle kit consists of a retaining spring wire support, a rubber block guide, and one universal-bridge metal frame spectacle front with mounting prongs. Corrective lenses are not included. The spectacle kit should be taken to a medical facility for installation of an individual's prescription lenses. Once the prescription lenses have been installed, the spectacle kit can only be used by that individual and shall be retained as personal glasses.

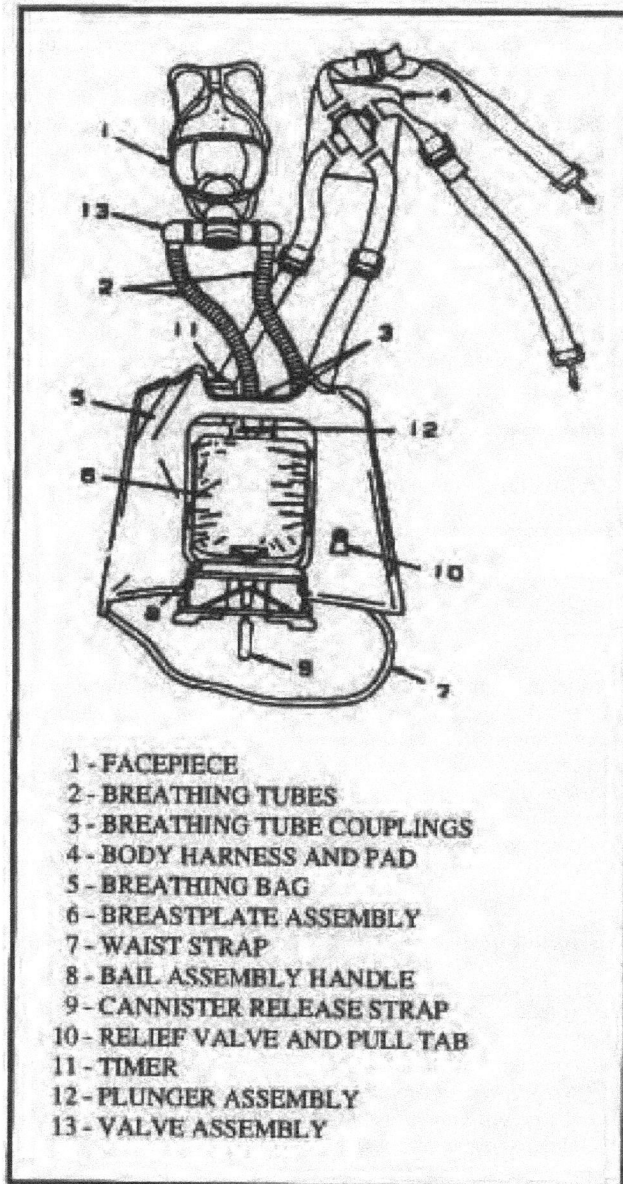

Figure 3-3. Navy type A4 OBA

1 - FACEPIECE
2 - BREATHING TUBES
3 - BREATHING TUBE COUPLINGS
4 - BODY HARNESS AND PAD
5 - BREATHING BAG
6 - BREASTPLATE ASSEMBLY
7 - WAIST STRAP
8 - BAIL ASSEMBLY HANDLE
9 - CANNISTER RELEASE STRAP
10 - RELIEF VALVE AND PULL TAB
11 - TIMER
12 - PLUNGER ASSEMBLY
13 - VALVE ASSEMBLY

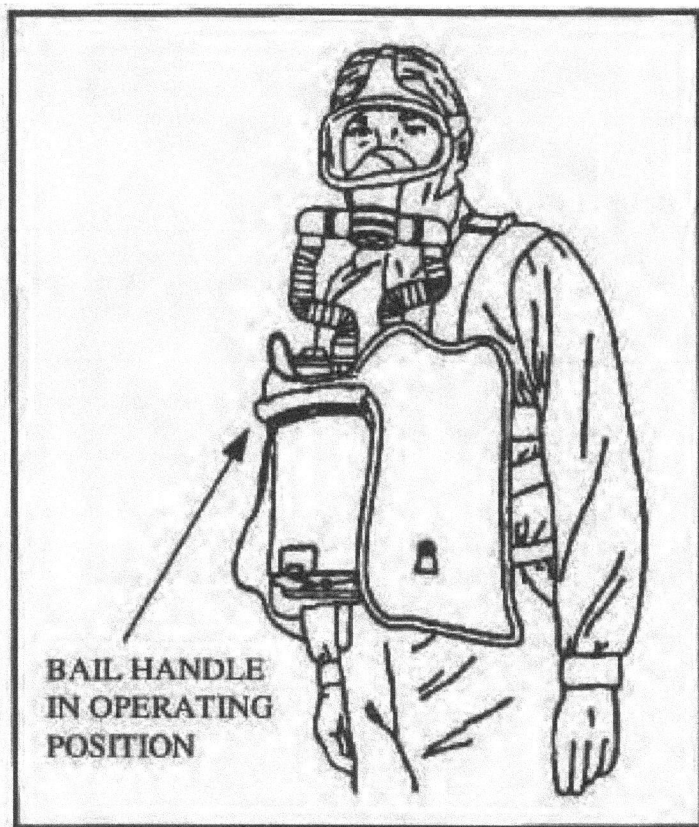

BAIL HANDLE
IN OPERATING
POSITION

Figure 3-4. Properly donned OBA

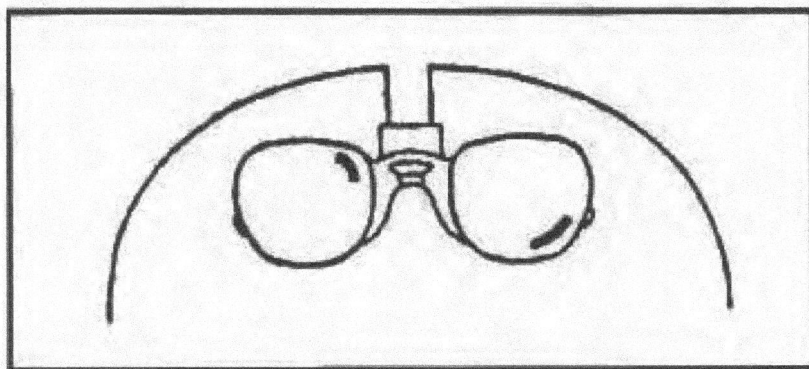

Figure 3-5. Spectacle Kit

OPERATION

3-102. It is necessary that the wearer be thoroughly familiar with the principles of operation and the procedures and precautions for using the OBA as a personnel protection device. The procedures for operation shall be practiced frequently under the supervision of a qualified operator. The user must don the OBA prior to entering a contaminated area. During the donning procedure, the canister will be inserted, the face piece donned and adjusted for proper seal, the bail handle raised to the operating position (figure 3-4), the canister fired, the OBA tested for proper operation, and the timer set.

DONNING AND ADJUSTING.

NOTE: While all of the donning and adjusting steps can be performed by the wearer, assistance will significantly speed up the process and provide additional assurance that the OBA has been properly donned and adjusted. However, since assistance may not always be available, wearers must be able to don and adjust the OBA without assistance.

- **Step 1** - Check that the bail handle is down and locked in the standby position. Do not raise the bail handle from the standby position to the operating position without first properly inserting a canister into the apparatus. Raising the bail handle without a properly inserted canister may damage both the plunger assembly and guide rods.
- **Step 2-** Attach face piece breathing tube quick-disconnect couplings (if unattached) as follows:
  - Fully retract the spring loaded outer sleeve of the coupling to expose the ball bearings.
  - Push the couplings firmly over the nipples (black onto black and blue onto blue).
  - Release the spring loaded outer sleeves.
  - To test the connection, grasp the hose at the clamp and pull lightly. If it is on correctly, the coupling will not pull off.

NOTE: Couplings are different sizes and are color coded to ensure proper assembly. It is possible to slide the larger coupling onto the smaller nipple, but it will not latch or seal. The OBA will not function if the couplings are not properly installed.

- **Step 3-** Fully extend and straighten all body harness and waist straps. Extend the head straps and place them in front of the face piece lens.
- **Step 4** - With one hand, grasp the face piece by the combination valve assembly and the apparatus by the bail handle. With the other hand, grasp the body straps by the body harness and pad. Bring the pad and harness over the head (figure 3-6) and position the OBA on the chest.

NOTE: DO NOT let the face piece hang down by the breathing tubes.

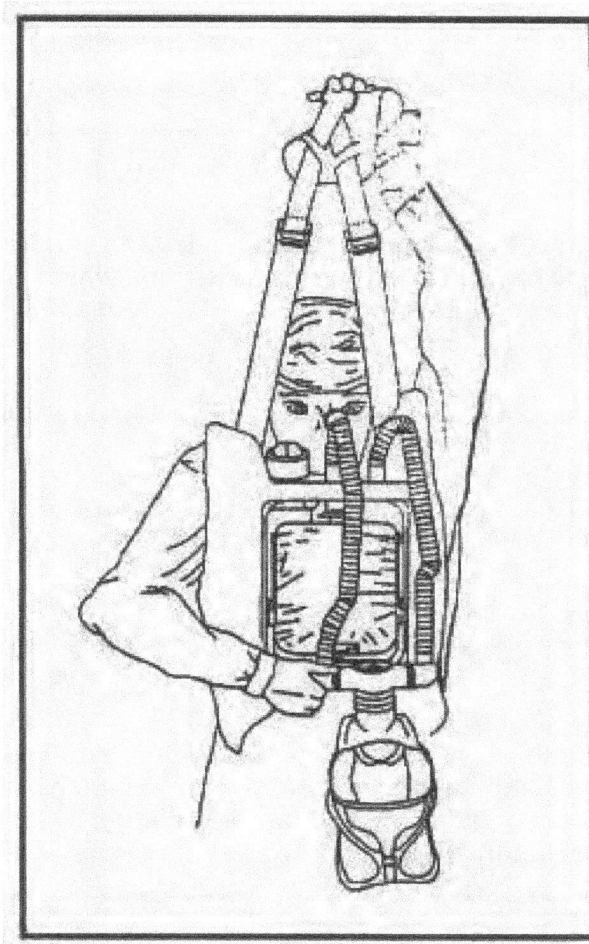

Figure 3-6. Lifting straps over head

- **Step 5 -** Run the underarm straps under the arms and attach the snap hooks to the rings on the top corners of the breastplate assembly.
- **Step 6 -** Position breast plate so the breathing tubes are slightly below the shoulders. While the apparatus is held in position, adjust the underarm straps, then the shoulder straps until it is fitting comfortably. When adjusted correctly, the harness pad should be located in the top center of the back and head movement will not be restricted after donning the face piece.
- **Step 7-** Place the face piece over the head so it is out of the way.
- **Step 8 -** Snap the waist strap to the bracket on the lower side corner of the breastplate. Adjust this strap to hold the apparatus snugly to the body. Wrap the excess strap under the secured part of the strap. If there is excess strap from the underarm straps, secure these under the waist strap.

NOTE: The waist strap can be quickly adjusted by changing the size of the waist strap loop. A nylon tie wrap attached to the loop will prevent the loop from being accidentally pulled out.

- **Step 9** - At this time, install the canister in accordance with the following procedure:
  - Remove the canister tear-off cap by pulling the tab backward and downward.
  - Remove the metal protective disk, exposing the copper foil seal and O-ring. Discard the cap and metal disk. Inspect the copper foil seal and O-ring to ensure both are intact. Do not puncture the copper foil seal.

WARNING

DO NOT pull the lanyard when removing the candle cover. Pulling the lanyard removes the cotter pin, which fires the candle and generates oxygen. The cotter pin and lanyard are shown in Figure 3-7.

  - Pull the swivel plate up and toward the center of the canister.
  - With the bail handle locked in the standby (down) position, insert the canister upward into the guard, with the neck up and the concave, ribbed side toward the body. The canister is correctly inserted when it is firmly retained by the canister retaining mechanism. This is called the standby position (figure 3-7).

WARNING

DO NOT use an OBA which pierces the foil seal in the standby position. If the copper foil seal is pierced when the canister is placed in the standby position (figure 3-7), adjust the standby stop.

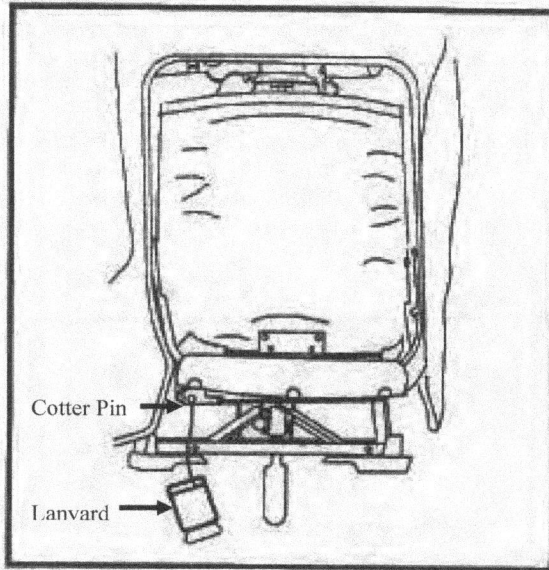

Figure 3-7. Canister held in standby position

> ### WARNING
>
> **If hair is allowed to penetrate the seal between the face and mask, it may result in a loss of oxygen from the BA and penetration of toxic fumes from the outside. The portion of the face in contact with the seal will be clean shaven to maintain an effective seal.**

- **Step 10 -** Don the face piece as follows:
  - Insert the chin into the chin stop in the face piece (figure 3-8).
  - Pull the head straps on front of the face piece over the head and make sure that the straps are lying flat (figure 3-9).
  - First tighten both lower straps at the same time (figure 3-10). Next tighten both upper straps at the same time (figure 3-11). Do not tighten the forehead strap at this time.
  - Place both hands on the head harness pad (on back of head) and push it down toward the neck.
  - Retighten the lower and then the upper straps.
  - Tighten the forehead strap if needed.

NOTE: When properly donned, both lower straps are tightened equally, both upper straps are tightened equally, the face piece is centered on the face, and the head harness pad is centered squarely on the back of the head.

- Test the face piece seal (figure 3-12). To do this, squeeze the corrugated breathing tubes together tightly with one hand and inhale gently. The face piece should collapse inward and remain collapsed while breath is held. This indicates there is a gas tight seal. Hold breath for at least five seconds. If a leak is detected, readjust the head harness straps. The face piece should be tested each time it is donned.
- Make final adjustments to all four body harness straps. The wearer should be able to look up and down without the face piece shifting or the breathing tubes catching on the timer.
- If going into standby or ready condition, loosen the lower face piece straps only. This allows the wearer to remove the face piece and place it over the head out of the way until needed.

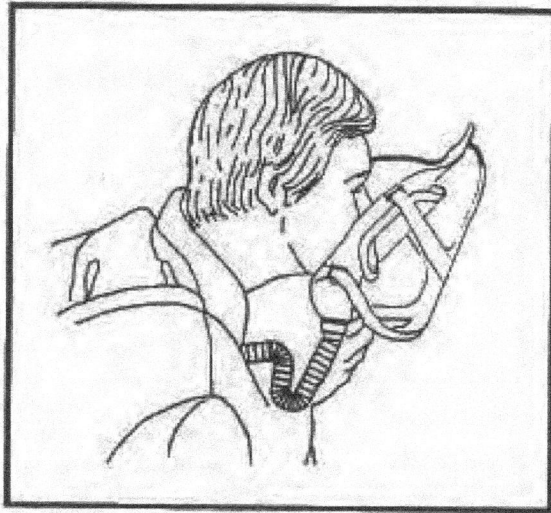

Figure 3-8. Inserting chin into face piece

Figure 3-9. Pulling straps over head

Figure 3-10. Tightening lower straps

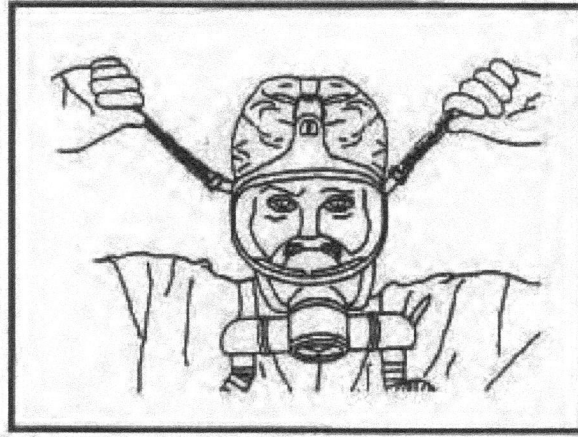

Figure 3-11. Tightening upper straps

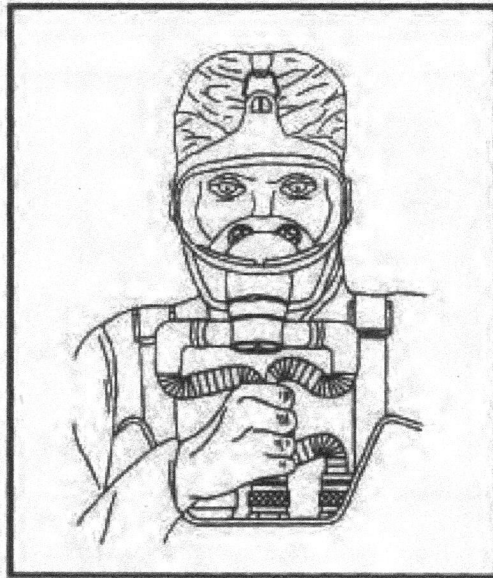

Figure 3-12. Testing face piece seal

PLACING OBA IN OPERATION

3-103.   After the OBA has been properly donned and adjusted, use the following steps to place it in operation:

- **Step 1 -** If in standby, don the face piece and retighten the lower straps. Then retest the seal of the face piece.

- **Step 2-** Using both hands; depress tabs on the bail handle to unlock the bail assembly from the bottom locked position. Swing the bail handle upward until it snaps into position. Push the bail handle forward without depressing the tabs. The handle should not move if locked in position.

NOTE: If the metal canister cap has not been removed, the bail handle will not swing upward into the operating position, and the plunger assembly may be damaged.

WARNING

DO NOT pull the breathing bag tab during normal operation. This will cause a loss of oxygen from the bag.

- **Step 3 -** Breathe normally. The exhausted breath will cause a chemical reaction, which will generate new oxygen from the canister. There will be more oxygen in the bag than is required. Excess oxygen will vent from the bag automatically through the relief valve in the bag. If the relief valve should stick after extended stowage time, use the breathing bag pull tab to activate the relief valve. While pulling the tab, check the breathing bag with the right hand to ensure bag does not deflate completely.
- **Step 4** – Conduct a leak test of the system. Check by feeling and listening for air leaks around the face piece, along hoses and oxygen bags while compressing the air hoses and bag firmly with the hand (figures 3-12 to 3-14).
-

WARNING

Once the timer bell has sounded, start leaving the contaminated atmosphere area and return to fresh air.

- **Step 5 -** Once the apparatus is inflated, working, and has been leak tested, set the timer. Rotate the pointer clockwise to 60 minutes and then set the time for 20 minutes. Rotating the time to 60 minutes before setting the timer to 20 minutes is required to completely wind the timer bell.

NOTE: The pointed end of the timer handle will be pointed directly away from the wearer when the bell sounds.

Figure 3-13. Testing OBA for leaks by compressing bag

WARNING

The leak test described in step 4 will be performed after each fresh canister is inserted and started. If the face piece is removed, leak test the face piece after putting it back on.

Canister Manual Starting

3-104. If there is a shortage of canisters and sufficient time, the canister may be started manually by performing the following steps:

- **Step 1** - Grasp both breathing tubes with one hand and squeeze tightly.
- **Step 2** - Work fingers from the other hand under the face piece to break the seal. Then inhale to draw external air into the face piece.
- **Step 3** - Remove fingers to reseal the face piece. Release the breathing tubes and exhale into the face piece. This forces air through the canister and into the breathing bags.
- **Step 4** - Repeat steps 1, 2, and 3 until the breathing bags are fully inflated.
- **Step 5** - Deflate the breathing bags by applying pressure on the right-hand side. Continue this procedure until the right bag is deflated.
- **Step 6** - Re-inflate and deflate the breathing bags as directed in steps 1 through 5.
- **Step 7** - Carefully feel the bottom of the canister without gloves. If it is warm, oxygen is being generated and the apparatus is ready for use. Set the timer and proceed with work. If the canister is not warm, repeat steps 1 through 5.

NOTE: Cold temperature and wind dissipate the heat generated by the chemical reaction and tend to slow it down. In lower temperatures, several cycles of inflating and deflating the bag may be required to start oxygen production.

Canister Removal

3-105.   The steps for canister removal are:

- **Step 1** - If the canister has been used, loosen the lower straps and remove the face piece. Put it over your head in the standby position. The face piece may be left on if another canister is going to be inserted into the OBA immediately.
- **Step 2-** Unlock the bail handle by depressing the tabs. Push the bail handle down from the operating position to the standby position. Loosen the waist strap, spread your legs apart, lean upper body slightly forward, and pull the canister release tab, while keeping hands away from the falling canister. The canister should drop out. If the canister fails to fall out, shake the OBA. This should free the canister.
- **Step 3-** If the canister stills fails to drop out of the apparatus, insert a thin metal rod between the inhalation and exhalation tubes. Then, pull the release tab and attempt to force the canister out. If this does not free the canister, set the OBA aside and allow the canister to cool. Then remove the canister using gloves.

WARNING

Expended canisters are hot and will burn unprotected skin. DO NOT attempt to touch the canister during removal without wearing fireman's gloves. In the event the expended canister cannot be disposed of after use and must be temporarily stowed until disposal action can be taken, extreme caution must be exercised when handling and stowing the expended canister. Care will be taken to prevent the entry of any foreign substances into expended canisters (particularly grease, oil, or water). Any of these substances can cause a violent chemical reaction and may even cause the canister to explode.

Canister Replacement

3-106.   Replacement of quick start canisters while the OBA is in use can be carried out indefinitely. Return to a non-contaminated atmosphere at the end of 20 minutes. Once in the non-contaminated atmosphere, replace the used canister with a fresh canister.

REMOVING OBA

3-107.   Perform the following steps to remove the OBA:

- **Step 1** - Remove the face piece by releasing the head straps at the buckles before pulling it off.
- **Step 2** - Place the face piece over your head in the standby position and remove the canister.
- **Step 3** - Loosen the waist strap and unhook it.

- **Step 4** - Loosen the shoulder straps and unhook the underarm straps from the upper corners of the breastplate assembly. Grasp the face piece and bail handle with one hand. Grasp the shoulder harness, preferably at the D-ring connector, with the other hand and lift the harness over your head.
- **Step 5** - Wipe down the OBA if it is wet or moist.
- **Step 6** - Clean the OBA after each use.

CANISTER DISPOSAL

> **WARNINGS**
>
> **Never handle opened canisters without suitable hand and eye protection (firemen's gloves and goggles). The canisters contain caustic chemicals that will injure the skin and eyes and should not be allowed to come in contact with the person. Never allow expended or unexpended chemicals to spill from the canister. These chemicals may cause combustion of any flammable materials with which they are brought into direct contact, especially if the materials are moist. Any spills will be cleaned up immediately and placed into a double wrapped poly bag, seal, and turned into proper authorities for at-shore disposal. Use a dustpan made of metal or nonflammable materials to cleanup spills.**
>
> **Use dry tools free of grease, oil, and water when puncturing canister copper foil seals.**

NOTE: If the canister tear-off cap has been removed, but the neck seal has not been damaged and the copper foil seal has not been punctured, the canister may be recapped for later use. Recap the canister with a new metal cap (NSN To be determined).

3-108. **Temporary Stowage of Defective or Used Canisters.** To dispose of defective or used canisters, perform the following steps:

- **Step 1** - Place unusable and unfired canisters in a clean metal bucket. Puncture the copper seal of the unusable or unfired canisters (if not already punctured), and then tire the candle. Set aside and let the canister produce oxygen from the candle for at least 15 minutes.
- **Step 2** - After cooling enough to be handled, recap canisters to be disposed of with a new metal cap.
- **Step 3** - Next, wrap the canister to be disposed of in double poly bags (NSN 8105-00-299-8532 or equivalent). Stow the poly bag wrapped canisters in a dry, oil-free environment until proper at-sea or shore site disposal is possible. Stowage will be carried out in a manner to protect against tearing and exposure to heat sources that could melt or ignite the bag.
- **Step 4** - If step 2 cannot be achieved, canisters will be stowed in sealed, clean, dry, oil-free metal containers. Containers will be of the open head closeable drum type with a gasket. Stow containers in a cool, oil-free space until proper shore site disposal is possible.
- **Step 5** - Upon arrival in port, contact the department ashore responsible for hazardous waste management. Arrange for off loading of used or unusable BA canisters.

NOTE: In an emergency, it may be necessary to dispose of expended canisters overboard. This is permitted only when the ship is outside of 25 nautical miles of shore. Care will be taken not to dispose of canisters in an oily environment.

PRECAUTIONS

3-109.   The following precautions are to be followed when using the OBA:

- Activate the OBA when ready to enter a compartment that has the possibility of a contaminated atmosphere.
- Prior to entering contaminated atmospheres, check the apparatus to make sure it is airtight in accordance with the donning and operating instructions
- Ensure that the breathing bags are properly inflated before entering the compartment. If excess time is needed to fill the bag or rapid deflation after filling occurs, conduct a leak check.
- Once the apparatus is inflated, set the timer.
- Turn the pointer on the timer dial clockwise to 60 minutes and then turn it back to 20 minutes. Rotating the time to 60 minutes prior to setting it to 20 minutes is required to fully wind the timer bell.
- Take care to protect breathing bags, breathing tubes, and face piece from damage. If any of these are torn or pierced while working in an unsafe atmosphere, cover damage with hand and return to fresh air immediately.
- If canister is changed in fresh air without removing the face piece, follow the canister starting and OBA leak checking procedures before leaving fresh air.
- Never release the face piece seal in an unsafe atmosphere, even if inhalation becomes difficult. Check the breathing tubes to see if they are kinked and restricting air flow. If a kinked tube is not the problem, return to fresh air immediately and have the OBA thoroughly checked.
- While in operation, check the timer knob frequently by feeling the pointed end. This is the best way to ensure that the timer is working and to check the remaining time, especially when there are high noise levels and poor visibility.
- When the timer bell sounds, the canister has been used for 20 minutes and the wearer will return to fresh air. The wearer will also return to fresh air if it becomes difficult to exhale or if the lens fogs up when inhaling. Remember, the approximate on-station time for an OBA canister in continuous use is 20 minutes.
- If hydrocarbon vapors are present, the OBA should not be used for more than a total exposure time of 3 hours (intermittent use only). After an exposure of 3 hours, rubber parts should be replaced. These parts include the following:
  - Face piece assembly and head harness.
  - Corrugated breathing tubes.
  - Breathing bags.

WARNING

DO NOT use an OBA for diving. Water entering through the face piece and exhalation tube will react violently with the chemicals in the canister.

3-110.   When properly fitted and operated, the OBA forms a closed breathing loop with the wearer's respiratory system. In an emergency, it can be worn in partially flooded compartments, however caution must be exercised. The danger of water entering through the seal at the canister neck is negligible; however, the danger of a violent chemical reaction due to water entering the canister through the face piece and exhalation tube must be considered and guarded against. Should the water level cover or partly cover the breathing bags, breathing will be more difficult. The added buoyancy at the wearer's chest will cause difficulty in balance and create a buoyant effect similar to that of a life jacket.

WARNING

Under no circumstances should the oxygen producing candle be saved for emergency retreat from the compartment. This practice is dangerous since candles have been known to misfire.

3-111. **Canister Precautions.** The following precautions are to be observed when using an BA canister:
- Canisters will be stowed in a cool dry place.
- Never stow canisters in the OBA.
- Do not remove the tear-off cap until ready to insert the canister into the OBA.
- Never remove and replace canisters in a contaminated atmosphere.
- Do not stow canisters near oil, water, or grease.
- Insert the canister into the OBA and lock the canister in the operating position
- Never try to reuse a canister. Once the copper foil seal is pierced and the canister has been removed from the OBA, the canister will be considered expended.
- Used canisters are very hot and should never be handled without suitable hand protection. Never allow any substance (particularly oil, water and oil mixtures, gasoline, or grease) to enter the neck of the canister. A violent reaction occurs when these substances come in contact with the oxygen producing chemicals. Never hold your face over the canister opening.
- Canisters should never be painted. Canister stock will be rotated to allow older canisters to be used first or for training. Canisters with paint peeling off should also be used first.

OBA TRAINING CANISTER KITS

3-112. The training canister contains a replaceable candle and removable firing mechanism. It is designed for use by 40 trainees. Each candle provides a five (5) minute supply of oxygen when fired. A new candle must be inserted into the canister for each trainee. Figure 3-15 is an exploded view of the training canister. Figure 3-30 illustrates the firing mechanism assembly. The following describes the use of training canisters. It also describes the stowage, precautions, and disposal of training canisters.

3-113. **Training Sessions.** All personnel will attend a training session on use the use of the OBA in accordance with commander instructions. All ships with OBAs have training kits as part of their allowance. The training kits will be used for classroom training of personnel in the use of the OBA. The training kits are used to ensure complete and realistic training for all personnel.

NOTE: Training with quick start canisters is highly recommended, if funding permits.

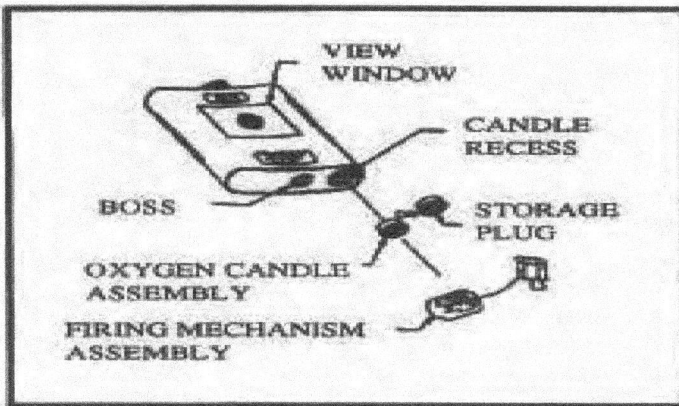

Figure 3-14. Exploded view of training canister assembly

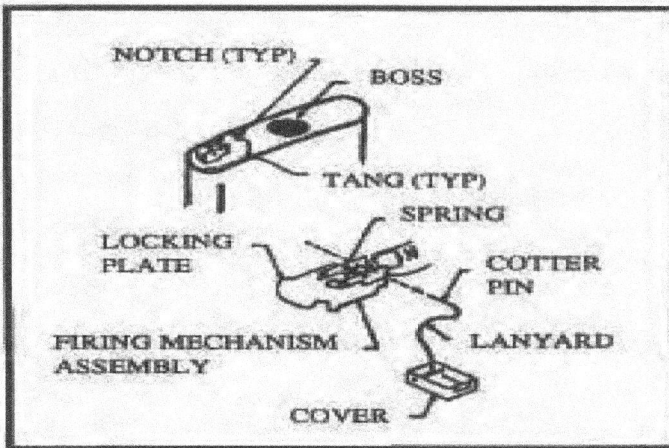

Figure 3-15. Training canister firing mechanism assembly

3-114. The following items are in each training canister kit:
- One red canister (quick start canisters are green)
- One storage plug
- Forty oxygen candle assemblies
- Forty tear-off caps
- One firing mechanism assembly (figure 3-16)
- One protective housing assembly
- Spare items include:
  - One storage plug
  - One firing mechanism assembly
  - Five protective housing assemblies

3-115. A window is provided on the training canister for checking the chemicals. If the view window on the canister changes from solid pink to a solid blue color, discard the canister. The training canister should be discarded if it has been used by 40 trainees (been fired 40 times).

3-116. **Oxygen Generation.** Oxygen used in the training canister is generated by a replaceable candle, which produces a five (5) minute supply (10 liters) of oxygen. Oxygen generation is started by firing the canister as you would a Manuel Start canister. Oxygen generation starts within 15 seconds after the candle has been fired. A small amount of harmless smoke may be present when initially fired. This is a normal condition and is a positive sign that the candle is working. Another sign signifying that the candle is generating oxygen will be when the breathing bags inflate.

3-117. **Charging Instructions for Initial Use of Training Canisters.** Perform the following charging steps prior to initial use of a training canister.

- **Step 1** – Do not remove the cover from the firing mechanism and then remove the mechanism and storage plug. Save the storage plug and reuse it when storing a partially used canister to prevent moisture from entering the canister body.
- **Step 2** - Insert the candle assembly into the candle recess in the canister. Ensure that the gasket is laying flat against the bottom of the canister.
- **Step 3** - Ensure that the firing mechanism is cocked and the cotter pin in place. Place the firing mechanism over the candle assembly with the candle primer housing projecting through the center hole. Notches should be lined up with tangs.

NOTE: Notches and tangs are positioned so the unit can be assembled only in the correct position.

- **Step 4** - Rotate firing mechanism clockwise to lock the candle in place. Tighten until the tangs are in contact with the firing mechanism frame. Do not over tighten.
- **Step 5** - Fold the lanyard into the candle cover. Secure the candle cover on the firing mechanism frame.
- The canister is now ready for use in the OBA.

3-118. **Charging Instructions for Previously Used Training Canisters.** Perform the following charging steps for canisters which have previously been fired.

- **Step1** - Secure the metal tear-off capon the canister neck. Ensure the hand tab is pointing toward the instruction label on the ribbed side of the canister.

WARNING

The firing mechanism and candle will be hot after use.
Wear gloves or allow the items to cool before handling.

- **Step 2** - Remove the firing mechanism from the bottom of the canister. To accomplish this, turn the firing mechanism counterclockwise until the tangs line up with the notches.
- **Step 3** - Remove and discard the used candle.
- **Step 4** - Hold the firing mechanism with the slotted end of the frame pointing toward you. Raise the firing hammer to the cocked position.
- **Step 5** - While holding the firing hammer in this position, replace the cotter pin through the holes in the frame from the same side.
- **Step 6** - Install the new candle and replace the firing mechanism in the training canister.

3-119. **Starting the Training Canister.** Perform the following steps for starting the training canister.

- **Step1** - Don and adjust the OBA in accordance with the steps contained in paragraph 31-9.
- **Step 2** - Remove the tear-off cap from the charged canister. Pull the tab straight backward and downward. Remove metal protective disk to expose the rubber O-ring.

- **Step 3** – Do not pull the lanyard.
- **Step 4** - Don, adjust, and leak test the face piece as outlined in paragraph 3-91, step 4.
- **Step 5** - With the bail handle locked in the standby position, insert the canister into the OBA with the concave, ribbed side toward your body. This is the standby position. Depress the bail handle tabs and swing the bail handle up until it locks in the operating position. Ensure the bail handle is locked in position by pushing forward on the handle without depressing the tabs. The bail handle should not move if it is properly locked in position.
- **Step 6** – Manually start the OBA canister by breathing into the face piece.
- **Step 7** - Set the timer on the OBA. Rotate the dial to 60 minutes and then back to 5 minutes.

NOTE: The setting for a quick start training canister is 5 minutes instead of 30 minutes.

- **Step 8** - After the training canister has been used for 5 minutes, remove the canister. To accomplish this, depress the tabs on the bail handle and swing the bail handle to the standby position. Loosen the waist strap, spread your legs apart, lean upper body slightly forward, and pull on the canister release tab while keeping hands away from the falling canister. The canister will drop out of the apparatus.
- **Step 9** - If the canister can be used for the next trainee, it may be recharged and stowed.
- **Step 10** - Following each use of the training canister, make a mark on the front of the canister with an indelible pen to indicate a use of the canister. When the canister has been used forty times or the canister view window changes from a solid pink color to solid blue, dispose of the canister as outlined in paragraph 3-113.

3-120. **Stowage.** Training canisters will be stowed in kits and locked in special lockers or in locked compartments. The kit is to be kept locked up at all times, unless being used for training. The Damage Control Officer will have control of the keys to ensure that training canisters will not be used during actual emergencies. After the training is complete or if a break of more than one hour is taken, partially used canisters will be sealed using the metal tear-off cap. Remove the oxygen candle and insert the plastic storage plug in the canister candle recess. The partially used canister will be returned to the special locker or compartment and locked up. Inspections of all components will be made to ensure that they are protected from moisture, which will cause deterioration. If lockers are not provided, stowage will be in a locked, dry, cool storeroom.

3-121. **Training Canister Precautions.** The following precautions will be observed when using the OBA training canister. This type of canister is for TRAINING USE ONLY. The training canister will never be used during actual emergencies. Simple exercises can be performed while the unit is in use for training purposes. It will be changed with a fresh candle prior to each use and will only be used in the presence of an instructor. Personnel experienced and qualified in the use of the BA will act as instructors. The use of the training canister by any trainee will never exceed 5 minutes. Prior to each use of the training canister, the instructor will inspect the color of the chemical through the view window. A pink color in the view window indicates a safe canister. As CO_2 is absorbed, the color in the view window will change to blue. When the canister has been used by forty trainees or if the color in the view window is completely blue (whichever occurs first), discard the canister. All trainees will be under the instructor's supervision at all times. The instructor will ensure that the breathing bags are inflated and continually supplying oxygen. If a low oxygen air mixture exists and is inhaled, near immediate collapse of the trainee will occur. Immediate aid will be rendered by the instructor. Training canisters will be kept free of oil, water and oil mixtures, gasoline, and grease. The training canister firing mechanism and candle assembly get hot after firing the candle. Exercise care when removing the training canister firing mechanism and candle assembly for recharging the canister. The chlorate candles used in the training canisters are subject to moisture deterioration. Therefore, training kits will not be opened until immediately prior to use. Keep kits closed whenever practicable. Do not remove candles from their protective cans until the canister is to be recharged.

NOTE: Only candles contained in individual cans with tear-off caps are allowed for use in training canisters. Older candles which are wrapped in plastic will be discarded.

3-122. Training canisters will not be stowed inside the OBA and training canister kits will not be stowed in damage control lockers. Do not attempt to cock the firing mechanism while the mechanism is mounted on the training canister. Do not remove the tear-off cap until the canister is to be inserted into the OBA. The canister will not be stowed without the tear-off cap and plastic storage plug in place. Never stow the canister with the oxygen candle installed in it.

3-123. **Disposal.** After the training canister has been used by 40 trainees or if the view window is a solid blue color, dispose of the canister. Once a chlorate oxygen candle has been burned, it contains harmless sodium chloride and partially oxidized iron and can be disposed of in the regular trash once it has cooled.

TENDING LINE

3-124. When the BA is in use, a tending line is provided for the user as a precautionary measure. The BA tending line is a 50-foot length of 3/16 inch aircraft cable with a clear plastic covering. The tending line has a stout hook at each end, which is closed with a snap catch. Either hook is used for attaching the line to the D-ring on the BA body harness pad. The cable is pliable and will slide freely around obstructions.

Use of Tending Line

3-125. When there is only one BA wearer in a compartment, a tending line will be used and there will be at least one additional person handling the line. The tending line is used to locate an injured person wearing the BA. To locate the injured person, follow the tending line to their location. Personnel tending the line should wear rubber gloves and boots. If there are two or more BA wearers in the same compartment, it is not necessary to use a tending line. BA wearers should keep in constant touch or sight of each other.

WARNING

If at all possible, stricken personnel will not be hauled by a line attached to their waist, or suspended from their waist. This can cause internal injuries. In an emergency situation, they may be dragged a short distance along the deck if no other means of rescue is possible. If the person lacks any sort of shoulder harness, make fast a line under their arms. Have the line meet either in front or in back.

3-126. Do not attempt to pull injured personnel out using the tending line. This may injure them even more, as well as costing valuable time. For a rescue to be effected immediately, rescue personnel standing by will have BAs donned and in the standby position. This will enable immediate entry when canisters are fired.

Tending Line Signals

3-127. When a tending line is employed, the wearer of the BA will remain in constant contact with the line tender. This is accomplished by the tending line signal system. Table 3-1 shows each signal and the corresponding meaning. The code for the signals spells OATH. This makes it simple to remember the signal system.

Table 3-1. Tending Line Signals

CODE	PULL	MEANING
O	1	OK
A	2	Advance
T	3	Take Up
H	4	Help

STOWAGE

3-128. **Stowage in Lockers.** Stow BAs in damage control lockers or in specified BA lockers located throughout the ship. Stow BAs in cool and dry areas of the damage control lockers. This prevents the build-up of moisture, which causes mildew damage to the rubber face piece, inhalation and exhalation tubes, and breathing bags. Keep all BAs away from oil, paint, and greasy substances. These are harmful to materials used in the construction of the BA. A properly stowed BA in a damage control locker is laying horizontal, one high, on a shelf with the face piece on top. A BA may be hung only in specific BA lockers.

3-129. **Stowage in Kit Bags.** BAs may be stored in the Firemen's Kit Bag along with up to two canisters, flash hood, flash gloves, fire fighting helmet, and fire fighting gloves. The kit bags may be hung on hooks which should be installed in such a way that the bags will be out of traffic and will be easily accessible. Care must be taken that the heavy body of the BA and other heavy items are placed in the bottom of the bag and the mask is placed on top to avoid damage to breathing tubes.

3-130. **Protection of Face piece.** The face piece may become scratched and damaged from entering and leaving tight spaces, while the mask is in the standby position. Damage may also occur from stowing and moving the BAs. To prevent damage and the need to replace the face piece, place a piece of surgical stockinet over the entire face piece down to the bottom of the breathing tubes. Cut the stockinet into three-foot strips. Knot each end and slip the open end over the face piece and breathing tubes. The stockinet should remain in place during stowage and when proceeding to the scene.

CLOTHING

> **WARNING**
>
> **The ensemble does not offer protection against chemical, biological, or radiological effects. The firefighter ensemble is intended to protect the firefighter from flame (flash) exposure, heat, and falling debris.**

3-131. The fire fighter ensemble consists of the following:
- Fire fighter's coveralls.
- Anti-flash hood.
- Anti-flash gloves.
- Damage control/firefighter's helmet.
- Fire fighter's gloves.
- Firemen's boots.
- Stowage bag.
- Flashlight, explosion proof.
- Breathing Apparatus (BA)

CONSTRUCTION

3-132. **Firefighter's Coveralls.** The coverall design is a one piece, jump suit style (Figure 3-17). The coverall has a tough outer shell, a vapor barrier, and an inner fire retardant thermal liner. The knees, bottoms of the thigh pockets, and bottoms of the legs are reinforced with leather padding for extra protection. As an additional safety feature, the coverall has reflective markings around the upper arms, lower legs, and torso to highlight the outline of the firefighter, so he can be seen in dense smoke or dim light. The front closure and inside lower legs have brass zippers. There are bellow pockets with Hook-and-loop closures on the outside of each thigh and on the front of the upper left arm. The coveralls have a corduroy faced collar with snap fasteners. The sleeves have an integral knit wristlet for wrist protection and small loops (thumb holes) on the ends of the sleeve wristlets to insert your thumbs to anchor and keep the sleeve from riding up the arm. The coveralls are available in five sizes (small through extra-large-tall).

3-133.   **Firefighter's Anti-Flash Hood.** The firefighter's anti-flash hood (Figure 3-18) provides protection to the head, neck, and face (except the eyes). The hood can be worn with the BA. It has an elastic face closure and is available in a single size which fits all. The face portion can be pulled up over the nose for additional protection of the face.

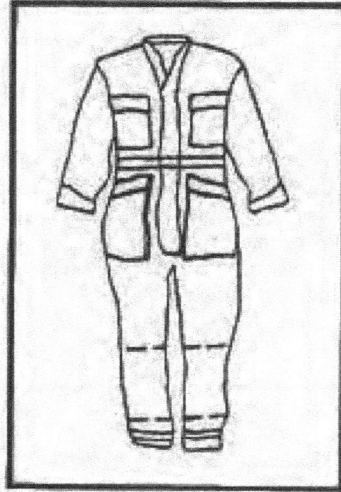

Figure 3-16. Firefighter's Coveralls

Figure 3-17. Firefighter's Anti-Flash Hood

3-134.   **Anti-Flash Gloves.** (Figure 3-19) The use of the gloves is to protect personnel from elevated air temperatures resulting in burns caused by fire. The gloves are made from fire retardant cotton and one size fits all.

3-135.   **Damage Control (DC)/Firefighter's Helmet.** (Figure 3-20) The helmet is designed to protect the head, neck, and face from flame (flash) exposure, heat, and falling objects. The helmet shell material is heat resistant fiberglass and is provided with a face shield, chin strap, adjustable suspension, reflective markings, and a liner that covers the side of the head and neck.

CAUTION

Do not modify the helmet in any manner, including removing the face shield and drilling holes to attach a light. Modification will reduce the protection provided by the helmet.

3-136. **Firefighter's Gloves.** (Figure 3-21) The gloves protect against abrasions, short duration flame (flash) exposure, and heat. The five-finger cut, gauntlet gloves are fabricated from leather and have a waterproof vapor barrier and fire retardant liner. The gauntlet provides wrist protection. The gloves are available in five sizes (extra small through extra large).

3-137. **Fireman's Boots.** (Figure 3-22) The rubber boots have steel safety toes and puncture proof steel insoles. Fireman's boots are available in two models, knee high and hip length. The US Army currently is using the knee high version. Knee high boots are worn inside the coveralls and are available in sizes 5 through 15.

3-138. **Stowage (Kit) Bag.** The stowage bag is provided to preassemble the ensemble for stowage. This bag is constructed of canvas duck with carrying straps. In emergency situations, after removing the ensemble from the bag, the stowage bag can be used for transporting other damage control equipment to the scene. In particular, the bag can be used to move spare BA's and canisters.

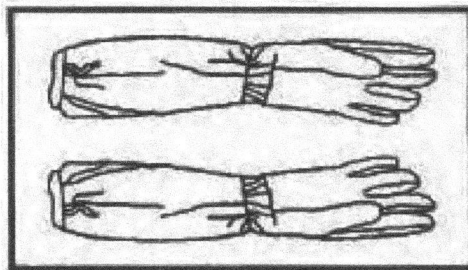

Figure 3-18. Anti-flash gloves

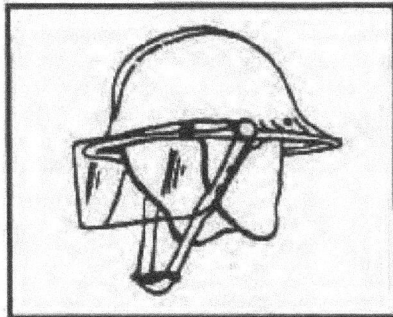

Figure 3-19 DC / Firefighter's helmet

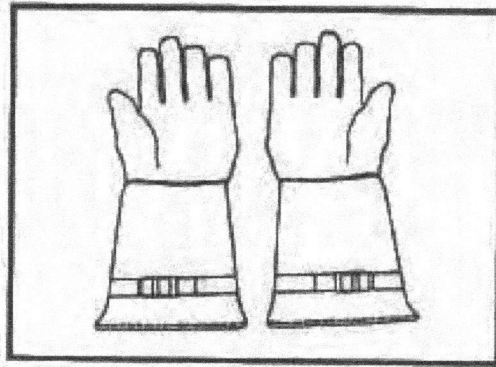

Figure 3-20. Firefighter's gloves

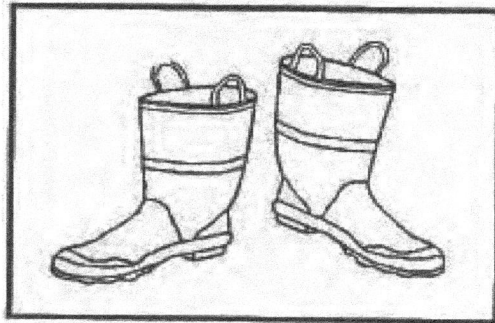

Figure 3-21. Fireman's boots

INSPECTION CRITERIA

3-139.   During emergency fire drills, inspect each piece of equipment prior to restowing. There should be no rips, tears, or holes in any gloves, coveralls, or hoods. Helmet should not be cracked or have any missing components. Helmet face shield should allow clear visibility by not being damaged with scratches. Zippers of coveralls should not be corroded by any exposure to the elements. Boots should not have any cracks or split seams on the soles and sides. Explosion proof flashlights should have batteries removed and stored in clear plastic sealable bag for future use. All components should be clean and dried before being restowed.

DONNING AND ADJUSTING

3-140.   An intricate part of the fire fighting ensemble is the type A-4 Breathing Apparatus (BA). The following steps will instruct the user on how to don and adjust the fire fighting ensemble (figure 3-23) including the BA:

- The BA is stowed with an anti-flash hood protecting the face piece lens. Remove the anti-flash hood from the BA face piece lens ring and put the hood over the face.
- Keep your pants and shirts on. Remove your shoes or boots and remove anything else that will interfere with donning the coverall, such as items in pockets.

- Put on the coveralls and pull them up and over your shoulders.
- Insert thumbs through the small loops on the ends of the sleeve wristlets to anchor and keep the sleeves over the wrists and under the gloves.
- Step into the fireman's boots. Never put on the boots before you put on the coveralls.
- Secure the two zippers on the bottom of the coverall legs.
- Stand up the coverall collar and ensure the anti-flash hood is fully inside the collar and down the chest as far as possible.
- Close the coverall front body zipper and the two collar snaps.
- Don the BA. Do not secure the face piece.
- Pull the anti-flash hood face opening down around your neck.
- Put on the BA face piece, tighten straps, and check for face piece straps, with the elastic opening over your face. Secure the hook-and-loop-closure on the coverall collar.
- Put the helmet on, secure helmet liner flap hook-and-loop-fastener, and fasten the chin strap.
- Loosen face shield fasteners on the sides of the helmet brim and rotate the face shield over the BA face piece to protect the breathing apparatus from debris and water.
- Remove the gloves from the leg pockets and put them on. Ensure they cover the coverall wristlets.

NOTE: Keep the BA breathing tubes outside the coverall and helmet liner flap.

REMOVAL OF GEAR

3-141. To takeoff the gear, reverse the donning order. Remove the gloves, pull up the helmet face shield, loosen the helmet liner flap hook-and-loop-fastener, take the helmet off, and open the coverall collar closure. Pull the anti-flash hood down around your neck and take off the BA face piece. Take off the BA, pull off the anti-flash hood, step out of the boots, and remove the coveralls.

STOWAGE

3-142. The firefighter ensemble should be stowed in the ensemble kit bag. The ensemble shall be preassembled and the bags located in, or near, Damage Control (DC) lockers so that they are easily accessible. Before stowing, ensure the ensemble is clean and dry. Stow the anti-flash hood over the face piece of the BA.

WARNING

DO NOT stow firefighter's protective clothing, BA, and BA canisters inside the vessel's super structure or engine room.

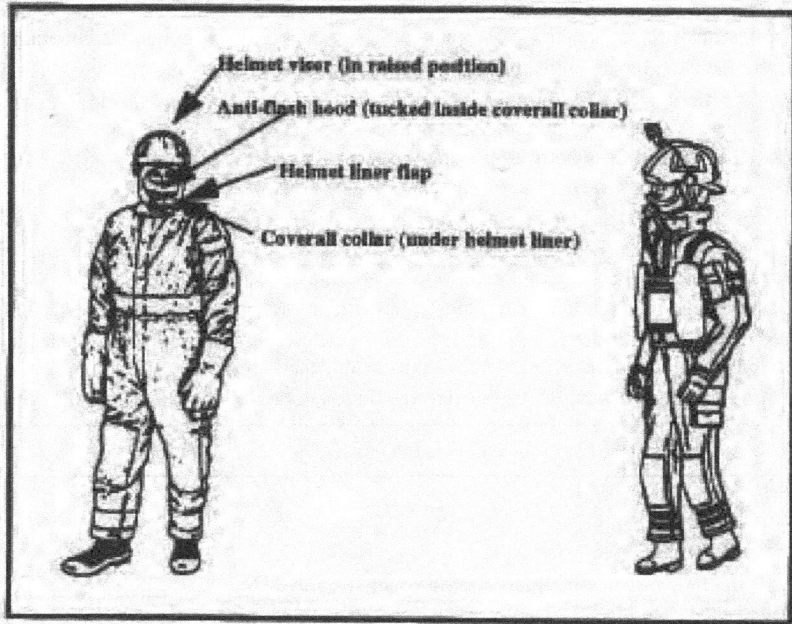

Figure 3-22. Fully donned ensemble with and without BA

HOSES

3-143. Fire hoses and nozzles shall be serviceable and connected to all fire stations. Fire hoses and nozzles will be maintained in accordance with this section. This section contains inspection and maintenance information about fire hose and fire hose nozzles used aboard Army watercraft.

VISUAL INSPECTION

3-144. Visual inspections will be made on fire hoses, nozzles, and hose couplings. The following describes the inspection of each of these items.

3-145. **Fire Hose.** Inspect each fire hose during weekly fire drills to determine that the hoses and nozzles are serviceable. Check to make sure that the fire hoses are free of debris. Inspect hose to ensure there is no evidence of mildew or rot, or damage by chemicals, burns, cuts, abrasions, and vermin. If the hose fails the visual inspection, it must be removed from service, destroyed, and replaced.

3-146. **Nozzles.** All nozzles will be inspected at weekly fire drills and after each use. Inspection will include the following:

- Clear of obstructions in waterway.
- No damage to tip.
- Tip chain is intact.
- Full operation of adjustments, such as pattern selection, and so on.
- Proper operation of shutoff valve.
- No parts missing.

NOTE: If the nozzle fails the inspection for any reason, it must be removed from service and repaired or replaced. Nozzles attached to in-service fire hoses will be kept in the closed position. If during use there is an obstruction that cannot be removed by flushing the nozzle, disconnect the nozzle from the hose and remove the obstruction through the hose connection end. Attempting to force the obstruction out through the tip may damage the nozzle. Handle nozzles with care. Avoid dents or nicks in nozzle tips, as this may seriously affect the reach of the stream. Nozzle control valves will be opened and closed slowly to reduce pressure surges. This would eliminate unnecessary strain on the hose and couplings. After use, all nozzles will be flushed and inspected before being placed back in service.

3-147. **Hose Couplings.** Couplings will be kept in serviceable condition. After use, and during each pressure test of the hose, they will be visually inspected for the following:

- Damaged threads
- Corrosion
- Slippage on the hose
- Out-of-round
- Swivel (not rotating freely)
- Missing lugs
- Other defects that impair operation
- Gasket for presence, tight fit, and deterioration

Couplings found defective will be removed from service and replaced. Do not drop couplings on steel deck or other hard surfaces. Doing this can cause damage to the threads. Do not allow vehicles to drive over couplings.

HOSE AND COUPLING PRESSURE TEST PROCEDURE

3-148. Fire hose and couplings will be tested annually to the maximum pressure they may be subjected in service, but not less than 100 pounds per square inch (PSI). Pressure tests may be performed by vessel's crew. Any length of hose that fails the visual inspection or service test will be removed from service and destroyed.

3-149. The following pressure test procedure will be followed:

- Total length of test hose line will not exceed 300 feet. The hose line shall be straight without kinks or twists.

WARNING

Questionable hose or hoses that have been repaired or re-coupled will be tested one length at a time.

- Connect the test hose line to a fire station valve. This valve must be manned during the test to prevent discharging a large volume of water in the event of a hose bursting during the test.
- Attach a nozzle to the far end of the hose line.
- With the fire station valve open and the end nozzle open, gradually increase the pressure to approximately 45 psi.
- Slowly close the end nozzle when the hose line is free of air and full of water.
- Close the fire station valve.
- Secure the hose line to avoid possible whipping or other uncontrolled reaction in the event of a hose burst.

WARNING

Clear all personnel from the area except those required to perform the remainder of the test procedure.

- Check hose line for leakage at the couplings. Tightened couplings with a spanner wrench where necessary.
- Mark each hose at the back of each coupling with a felt tip marker to determine if the coupling moves on the hose during the test.
- Slowly increase the pressure to test pressure (not less than 100 PSI) and hold for five minutes.
- Inspect for leaks, bubbles, and separation from ends while the hose line is at the test pressure.

WARNING

Personnel shall never stand in front of the free end of the hose, within 15 feet to the side of the hose, or straddle a hose during the pressure test.

- If a section of the hose is leaking or bursts, terminate the test. Drain the hose line and remove and destroy the failed hose.
- After the five minute pressure test, shutdown the pump, open the end nozzle to relieve the pressure, and drain the hose line.
- Observe the marks placed on the hose at the back of the couplings. If the coupling has slipped or twisted, the hose has failed the test. Remove and destroy the failed hose.
- Enter the test results in the ship's log.
- All hoses shall be cleaned, drained, and dried before being placed in service or storage.

MARKING

3-150. All fire hoses shall be marked with the vessel's name or number, test date, and test pressure.

NOTE: Do not replace old fire hoses and nozzles unless they are damaged or are no longer serviceable

---

Chapter 4

DAMAGE CONTROL EQUIPMENT

INTRODUCTION

Damage control procedures and equipment are covered in this Chapter. When an emergency occurs at sea, the crew must know how to save the vessel. Many times, it is not your own vessel you are saving, but another vessel in distress. Ability to put damage control knowledge to use without hesitation can mean the difference between life and death at sea.

There are three basic objectives of shipboard damage control which can be defined as occurring before, during, or after the ship sustains damage. Briefly stated, these objectives are as follows:

- Take preliminary action before damage occurs. This includes maintenance of watertight and airtight integrity, preservation of reserve buoyancy and stability, removal of fire hazards, preventive maintenance and test of emergency and damage control fittings and equipment.
- After damage has occurred, the focus must change in order to minimize, contain and localize casualty damage by controlling flooding, dewatering, firefighting, and transport and care of injured crewmembers.
- Lastly, to accomplish repairs and recover from the casualty as soon as possible after sustaining damage. This is accomplished by segregating ruptured piping systems, reconfiguration, establishing casualty power, and regaining a safe margin of stability and buoyancy by re-enforcing shored and weakened structures and manning essential equipment, as well as posting the appropriate watches (shoring, re-flash, etc.).

DEWATERING PUMP

P-100 PUMP UNIT

4-1.   The P-100 pump unit is commercial diesel driven portable pump designed for firefighting, dewatering, and many utility functions. The design features of the pump unit are described in the following paragraphs. The pump unit consists of the engine, centrifugal pump, exhaust primer, discharge valve, recoil starter, attached 1.45 gallon fuel tank, and compound pressure gage (figure 4-1). The pump unit measures 21″W X 23.5″L X 24.38″H. The wet weight of the pump unit is 164 pounds which includes 1.45 gallons of fuel that will allow 2.75 hours of operation. The pump is designed to provide 100 GPM at 83 PSI while lifting 20 feet. In high lift operations, the pump unit will deliver 68 GPM at 45 PSI while lifting 39 feet.

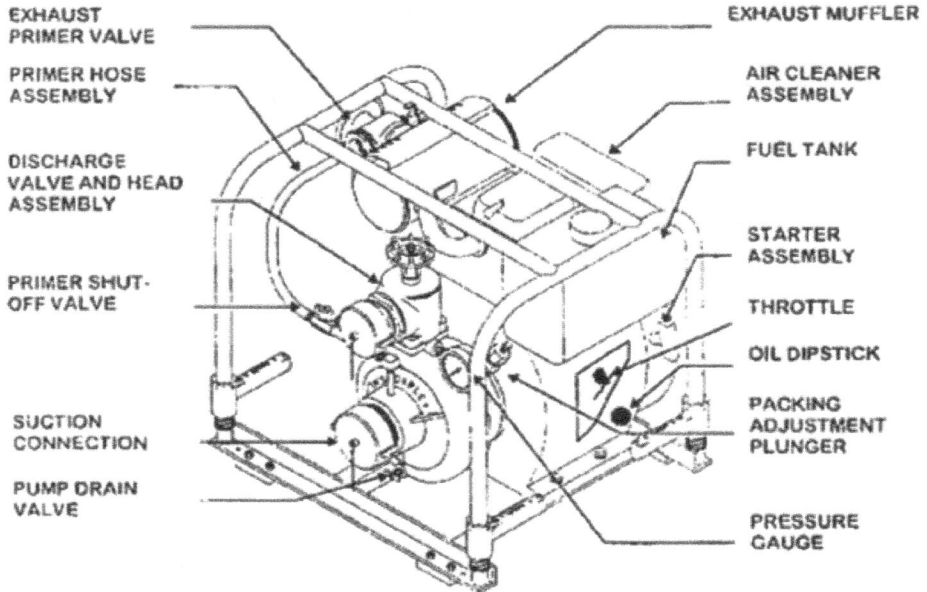

EXHAUST PRIMER VALVE

PRIMER HOSE ASSEMBLY

DISCHARGE VALVE AND HEAD ASSEMBLY

PRIMER SHUT-OFF VALVE

SUCTION CONNECTION

PUMP DRAIN VALVE

EXHAUST MUFFLER

AIR CLEANER ASSEMBLY

FUEL TANK

STARTER ASSEMBLY

THROTTLE

OIL DIPSTICK

PACKING ADJUSTMENT PLUNGER

PRESSURE GAUGE

Figure 4.1. P-100 Pump

ENGINE

NOTE: YANMAR ENGINE MODEL L100AE-D WAS SUPPLIED PRIOR TO JANUARY 2000. YANMAR ENGINE MODEL L100EE-D IS SUPPLIED AFTER JANUARY 2000.

4-2.   The Yanmar L100AE and L100EE engines are air cooled, single cylinder, four cycle diesel engine rated at 10 horsepower. Ignition is achieved by direct injection of fuel and compression is initially aided by a compression release lever to help overcome the 19.3 compression ratio. The engine is started by a recoil type starter. The engine's single cylinder has a displacement of 0.406 liters(24.78 cubic inches) which corresponds to the stroke X bore of 1-86 X 70 mm (3.386 X 2.756 in.).

4-3.   The fuel injection pump is a Bosch type Yanmar PFE-M type, timed at 13 plus or minus 1 b TDC. It supplies a Yanmar YDLLA-P type fuel injection nozzle which delivers fuel at an injection pressure of 19.6 Mpa (200 kgf/square cm). The fuel oil filter is a paper element type built into the 5.5 liter (1.45 gallons) attached fuel tank.

4-4.   The engine utilizes forced lubrication via trochoid pump and splash lubrication for valve rocker arm chamber. The lubricating oil filter is a resin, 60 mesh type. The engine lubricating oil capacity is 1.65 liters (0.44 gallons). The recommended oil for commercial use is SAE 10W30, API grade CC or higher for ambient temperatures less than 85 degrees F. The oil specified is MIL-L-2104, equivalent to SAE 15W40. The air cleaner element is a dry paper element type. The engine is cooled by forced air generated by a flywheel fan. Speed control is accomplished by an all speed type mechanical governor.

4-5.   The engine dimensions, length, width, height, are 417 X 470 X 494 mm (16.417 X 18.504 x 19.449 inches). The dry weight of the engine is 48.5 kg (106.9 lb).

PUMP

4-6. The Darley 2BE pump is a single suction, single stage centrifugal pump complete with a compound pressure gage, drain valve, and primer connection. The impeller is a closed design and the shaft is sealed by a unique palletized packing gland. The shaft seal utilizes injection plastallic packing with a stuffing box. The suction and discharge connections have male threads which receive 3 inch and 2-1/2 inch hoses, respectively.

4-7. The pump casing is fabricated from a hard coat anodized aluminum alloy which is light weight and corrosion resistant. The impeller is dynamically balanced and is of a bronze alloy construction. The wearing rings are a bronze labyrinth type.

EXHAUST PRIMER

4-8. The engine exhaust silencer is constructed to incorporate a jet type ejector and receive an insulated exhaust hose. When the primer is operated, the main exhaust port is blocked by the cylinder valve forcing the exhaust flow through the priming jet. The vacuum developed by the exhaust jet evacuates the air from the pump casing and suction hose. Because of the vacuum developed, atmospheric pressure forces water up through the suction hose and into the pump casing.

4-9. The exhaust hose is a dry 4.5″ insulated hose which is available in 10' sections. The hose weighs only 1.7 pounds per linear foot and provides adequate protection for safe handling with firefighter's gloves during and after operation. The function of the exhaust hose is to safely route harmful exhaust gases to weather when indoor operation becomes necessary.

FUEL TANK

4-10. The 1.45 gallon capacity fuel tank is mounted to the engine. The tank consists of the tank, fuel filter, isolation valve, injection valve, level sight tube, and a fuel tank cap.

PREPARATION FOR PRIMING

4-11. Check coupling gaskets and connect hose lines with couplings properly tightened.

4-12. A strainer with openings not larger than 1/4″ mesh must always be used on the end of suction line when pumping water from draft.

CAUTION

THE SUCTION HOSE MAY REQUIRE SUPPORT TO PREVENT EXCESSIVE WEIGHT FROM STRESSING THE PUMP CASING, INBOARD HEAD, OR ENGINE. WHERE PRACTICAL, THE SUCTION HOSE SHOULD BE TIED TO SOME NEARBY STRUCTURE AND/OR BLOCKS SHOULD BE PLACED BENEATH THE SUCTION HOSE ADJACENT TO THE UNIT TO RELIEVE STRESS ON THE PUMP.

4-13. Avoid air traps in suction hose if possible.

NOTE: BE CERTAIN THAT THE SUCTION HOSE (OR PIPE) IS ABSOLUTELY AIR TIGHT. NEITHER THE PUMP NOR THE PRIMER WILL LIFT WATER IF THE SUCTION SIDE OF THE PUMP HAS THE SLIGHTEST AIR LEAK.

4-14. Keep the suction intake strainer well above the bottom of the water source to prevent picking up soil and other foreign matter. If the strainer must lie on the bottom, a metal plate or pan should be laid under it.

NOTE: WATER MAY BE DRAFTED FROM PONDS, LAKES, STREAMS, CISTERNS, TANKS, SEA WATER, AND/OR WELLS. WHATEVER THE SOURCE, THE STATIC LIFT MUST NOT EXCEED 22 FEET FROM THE CENTER OF THE PUMP TO THE SURFACE OF THE WATER. A LIFT NOT EXCEEDING 10 FEET IS RECOMMENDED. THE SOURCE OF SUPPLY SHOULD BE REASONABLY CLEAR AND FREE FROM FOREIGN MATTER.

4-15. Submerge the suction intake sufficiently into the water to prevent sucking in air. A cover laid over the top of the strainer will allow the pump to operate with a minimum of submergence.

4-16. Close drain valve and all other openings into pump casing.

4-17. Do not start the engine until everything is ready for pumping, with hose couplings properly tightened. Pump discharge check valve may be partly open during priming at lifts less than 10 feet, and completely closed for lifts of 10 feet and more.

STARTING AND PRIMING THE PUMP UNIT

WARNING

DO NOT OPERATE THE PUMP UNIT IN CONFINED SPACES UNLESS THE EXHAUST HOSE IS CONNECTED TO CARRY THE TOXIC ENGINE EXHAUST GASES TO WEATHER.

WARNING

HEARING PROTECTION IS REQUIRED IN THE IMMEDIATE AREA OF THE PUMP UNIT WHILE IN OPERATION.

4-18. To start and prime the P-100 (figure 4-2):
- Set the fuel tank isolation valve located under the fuel tank to "O" (open) position.
- Set the engine throttle control to the "START" position.
- Open the primer line shut-off valve between the primer jet, and the pump suction. (Valve is open when knob is in line with the air passage.)
- Slowly pull on the recoil starter checking engine and pump for freedom of movement and priming the engine with lubricating oil. Depress the compression release lever ensuring that it remains depressed. The compression release lever will spring shut when the engine rotates during starting attempts.
- Start the engine by pulling the recoil starter rope (figure 4-3).

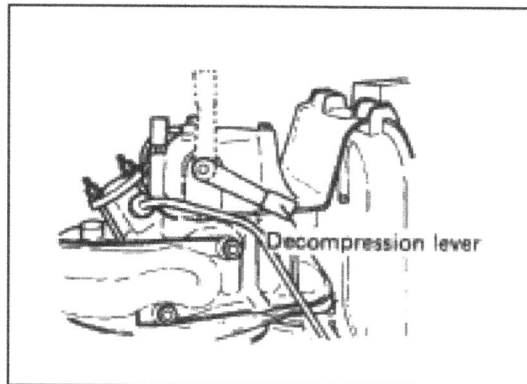

Figure 4-2 Starting and priming the pump unit

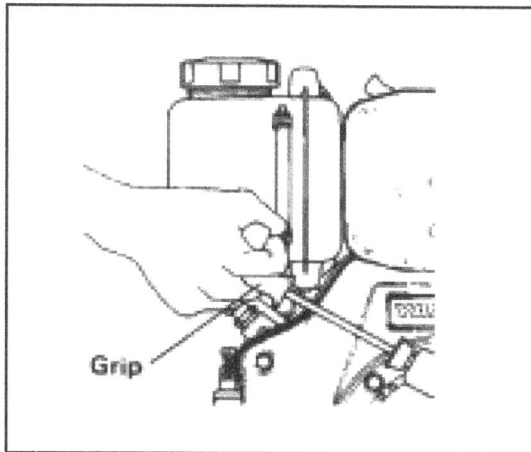

Figure 4-3 Starting the engine

- Once the engine is running, set the engine throttle control to the "RUN" position.

CAUTION

NEVER RUN THE PUMP AT HIGH SPEEDS, UNLESS IT IS DISCHARGING WATER.

CAUTION

NEVER RUN THE PUMP WITHOUT WATER ANY LONGER THAN THE SHORT TIME REQUIRED FOR PRIMING.

NOTE: START THE ENGINE AND RUN AT A FAST IDLE TO PRIME WITH LIFTS LESS THAN 10 FEET. START THE ENGINE AND RUN AT FULL THROTTLE TO PRIME WITH 10 TO 22 FOOT LIFTS.

● Shift the exhaust valve to the prime position blocking the main exhaust opening. The exhaust valve is in the prime position when the handle is horizontal.

NOTE: WHEN PRIMING ON HIGH LIFTS, OR WHEN PUMPING DIRTY WATER, IT MAY BE NECESSARY TO SEAT THE DISCHARGE STOP-CHECK VALVE BY TIGHTENING DOWN GENTLY WITH THE HANDWHEEL. UNSCREW THE HANDWHEEL WHEN WATER IS DISCHARGED THROUGH THE EXHAUST JET.

● When a steady stream of water appears at the discharge of the priming jet, close the primer line shut-off valve and return the engine exhaust valve to the normal position. Open the pump discharge valve.
● Repeat the priming operation if the pump fails to hold its prime. If the pump does not deliver water within two minutes, stop the engine and check for air leaks at suction connections and/or the pump packing gland, or failure of the priming jet to produce vacuum.

CAUTION

EXTENDED OPERATION WITHOUT PRIME MAY CAUSE SERIOUS DAMAGE TO THE PACKING GLAND, THE PUMP SHAFT, AND OTHER PUMP INTERNALS.

● After priming the pump with water, start the pump and raise the discharge pressure to 50 psi. Tighten the packing screw using a 6″ long 9/16″ end wrench until drip rate is between 5 and 60 drops per minute – do not over-torque (24 in-lb torque). Continue operating the pump at 50 psi for 5 minutes to dissipate packing pressure against the shaft and permit cooling water to flow between the shaft and stuffing box hole. Make sure that water actually does come through before operating pump at any higher pressure. The normal drip rate may vary between 5 and 60 drops per minute.

NOTE: THE PACKING GLAND SCREW SHOULD BE KEPT SUFFICIENTLY TIGHT TO PREVENT EXCESSIVE LEAKAGE ONLY. SLIGHT LEAKAGE IS ALWAYS REQUIRED DURING OPERATION TO COOL THE PACKING AND PREVENT DAMAGE TO THE IMPELLER SHAFT.

● Operate the pump for 10 minutes at the highest normal operating pressure flowing sufficient water to prevent overheating. Do not run the pump blocked tight. Lower discharge pressure to 50 psi and repeat the packing screw tightening procedure outlined above.
● The pump may now be operated for any time period required within its rated capacity, however, the drip rate should be monitored more frequently during the first few hours, and adjusted if necessary to achieve a stable flow rate. Several more adjustments may be required. All diesel engines must be throttled back by the operator in high load situations. This must be done to prevent over-fueling the engine as is evident by black exhaust smoke. Careful readjustment of the throttle will not cause a decrease in pump performance. Throttle back until pump performance just begins to decrease.
● While the pump unit is running, occasionally monitor the pump discharge gage and fuel tank level. For periods of extended operation, refueling may be necessary. Extreme caution is required when refueling a hot or running engine. An additional crew member must standby with an appropriate fire extinguisher should fuel inadvertently be spilled on hot engine parts.

> ### CAUTION
> OVER-FUELING THE ENGINE WILL CAUSE DILUTION OF THE ENGINE OIL AND PREMATURE WEAR ON THE CYLINDER WALLS AND BEARINGS.

SHUTDOWN

4-19. To stop the pump unit, reduce engine speed to an idle speed and allow the engine to cool down for two minutes. Return the engine throttle control to the "STOP" position. If engine continues to run, shut the fuel tank isolation valve. After Operating the P-100 Pump, if the pump has been used to pump seawater, the seawater must be drained from the pump by opening the pump casing drain valve. The pump must be flushed with fresh water to prevent corrosion and salt crystals from forming on close tolerance pump internals. After flushing the pump, apply a spray silicone compound to pump internals while slowly pulling the starter rope and replace hose connection caps. Drain water out of pump casing immediately. The drain valve is located at the lowest point in the pump casing. Do not forget to close all drain cocks after draining all water. If forgotten, trouble in priming will follow on the next run. Check lubrication after every run. Periodically inspect and run pumps used for fire service to ensure that they will be ready in an emergency.

HIGH SUCTION LIFT OPERATION

4-20. Install Vita Motivator eductor with foot valve and strainer on the submerged end of the suction line. The suction line must slope down all the way from the pump to the water. Hand priming with the Vita Motivator eductor can be easily achieved by filling the hose through a Y-gate valve or Tri-gate valve connected to the 1-1/2" feed line. By filling through the feed line, the check valve does not have to be held open as the water from the feed line will fill up the suction line and open the check valve.

COLD WEATHER OPERATION

4-21. The first assurance against cold weather trouble is to keep fire apparatus stored in heated quarters. When setting up for pumping, avoid unnecessary delays by thoroughly training pump operators. Be sure that primer lines are kept closed until ready for use. Have discharge lines ready so that pumping may be started as soon as it is primed. Do not stop flow of water through the pump until ready to drain and return to the station. Eliminate all water from pump casing and primer line between periods of operation.

> ### CAUTION
> IN COLD WEATHER, IT IS IMPORTANT TO MAKE SURE THE TUBING LEADING FROM THE EXHAUST PRIMER TO THE PUMP CASING IS FREE FROM WATER TO PREVENT FREEZING. FREEZING OF THIS TUBING WILL RENDER THE EXHAUST PRIMER INOPERATIVE AND MAY DAMAGE TUBING AND FITTINGS.

4-22. To remove the water from the primer tubing:
- Restart the engine after disconnecting the suction line.
- Open primer line shut-off valve.
- Close engine exhaust valve tightly with lever at the side of exhaust primer.
- After five seconds, open exhaust valve.
- Shut off engine.

TESTING EQUIPMENT FOR PRACTICE

4-23. Frequently, operators of a fire apparatus, who are not thoroughly familiar with its operation become confused under the stress of the emergency and neglect details that may cause trouble or delay in getting the equipment into operation. In light of that fact, practice tests be conducted repeatedly until operators are thoroughly trained. More than one individual in each department should be a competent operator. Practice should include pumping from low lifts, high lifts with short and long suction lines, with suction line elevated to form an air trap, from hydrants, and at large and small capacities.

CAUTION

NEVER BREAK OR RESTRICT SUCTION OR ADMIT AIR TO SUCTION LINE WHILE ENGINE IS OPERATING WITH THROTTLE OPEN. THIS WILL RELEASE THE LOAD AND POSSIBLY ALLOW THE ENGINE TO OVER-SPEED.

4-24. It is a good idea to note the effects of air leaks in hose, insufficient submergence and restriction of suction line. (Suction lines can be restricted by placing a can or other strong closure around the suction strainer.)

ELECTRICAL CENTRIFUGAL, SUBMERSIBLE PUMP

4-25. Ocean going Army vessels are authorized electrical submersible dewatering pumps. These pumps are small in size and must be configured to perform their function. The electrical cable must never be allowed to support the weight of the pump. A lifting line must be attached to the upper pump handle and be sufficiently long to have the electrical cable, with slack between each support, woven with half hitches from the lifting line approximately every three (3) feet the complete length in order to raise and lower the pump. Additionally, the pump discharge piping must have a suitable hose attached to allow the water being pumped to discharge outside the space being pumped.

FAN, WATER DRIVEN

DESCRIPTION

4-26. This portable water-driven fan (figure 4-4) is a high velocity blower that provides 2,500 cfm with 10" ducts and powered by a high speed water turbine. The turbine is driven by water from the ship's fire main, shore side fire main, or the P-100 emergency fire pump through a standard 1-1/2" fire hose. The inlet water coupling is equipped with a wire mesh strainer to prevent large particles from obstructing water flow to the turbine (Replacement for Red Devil Blower).

4-27. **Use.** This fan is used for emergency ventilation and de-smoking and bulkhead cooling while maintaining fire boundaries. Discharge water is directed overboard through navy standard 1-1/2" fire hose via an overboard discharge connection or by directing the fire hose over the side. The water-driven fan will operate with inlet water pressure from 50 to 180 PSIG, and its performance still varies according to the inlet water pressure and the discharge water back pressure. As in any turbine-driven piece of equipment, it is the pressure drop across the turbine that provides the power to drive the equipment. Low inlet water pressure, clogged inlet straining, or a high back pressure on the discharge water hose will reduce the fan's performance.

NOTE: The strainer **must** be inspected and cleaned to insure proper operation of the water driven fan.

4-28. Care must be taken when running the supply and discharge water hoses, as kinks in either hose can adversely affect the fan's performance. It is approved for use in explosive atmospheres. A 10" x 15' vent hose is used with the water-driven fan. Use adapter NSN 4730-01-378-5288 for connecting the vent hose. 50 psig water pressure will move about 700 cfm of air. 180 psig water pressures will provide optimum air movement of 2500 cfm. Also may be used with mister attachment.

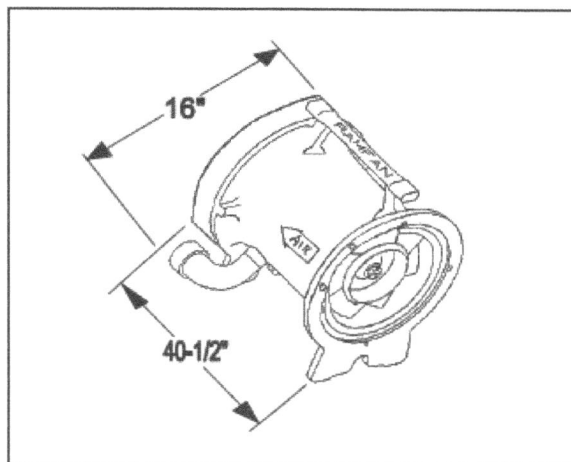

Figure 4-4. Water Driven Fan

WEDGES

4-29. The tapered wedges are made of hardwood and are used in shoring and plugging operations for pipe and hull repairs. They are available in many sizes. See figure 4-6 for measuring specifications. Wedges are authorized by the vessel's basic issue items (BII) and will be maintained in a dry location as specified by the vessels damage control plan.

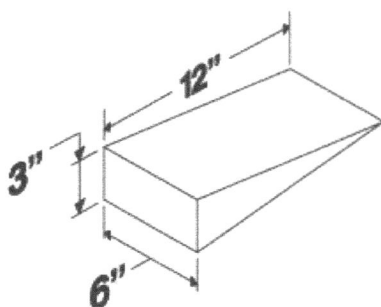

Figure 4-5. Example of how to measure a wedge for identification, H (3") x W(6") x L (12")

PLUGS

4-30. The conical tapered plugs are made of soft wood and are used in patching and plugging operations for pipe and hull repairs. They are available in four sizes: 1 x 0 x 3 inches, 2 x 0 x 4 inches, 3 x 0 x 8 inches and 5 x 1 x 10 inches (Figure 4-7) Plugs are authorized by the vessels BII and will be maintained in a dry location as specified by the vessels damage control plan.

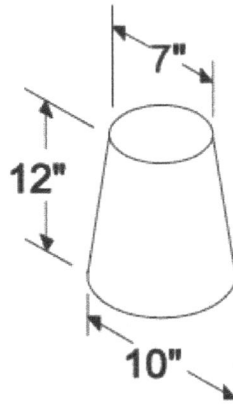

Figure 4-6. Example of how to measure a plug for identification, End Width (10") x Point Width (7") x Length (12")

EMERGENCY DAMAGE PIPE REPAIR KIT

4-31. **Soft Patch Kit.** Piping system leaks usually accompany any large hole in the hull. Soft patches can seal holes and cracks in low-pressure lines and water lines. Install a soft patch on a pipe as follows (figure 4-7):

- Plug the opening with soft wood plugs or wedges (the flow of water must not be retarded by driving an excessive amount of wood into the pipe).
- Trim plugs and wedges flush with the outside of the pipe.
- Wrap rubber sheeting over the damaged area and back it with light sheet metal held in place with bindings of wire or marline.
- Stop minor pipe leaks with a jubilee patch (an adjustable strap with a flange on each edge). These can be made by bending sheet metal around a cylinder and turning out the flanges and then clamped in place (figure 4-8). The flanges may have to be reinforced as pressure increases.

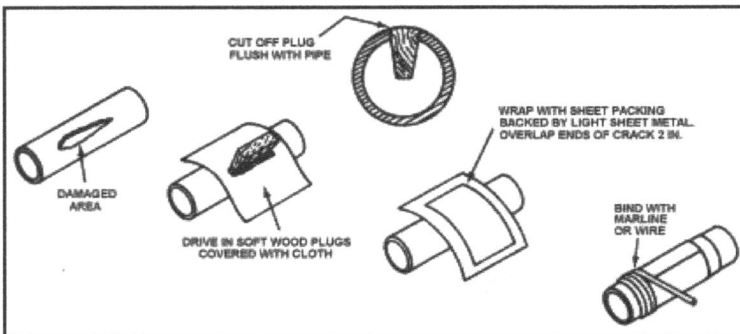

Figure 4-7. Installing a Soft Patch on a Pipe

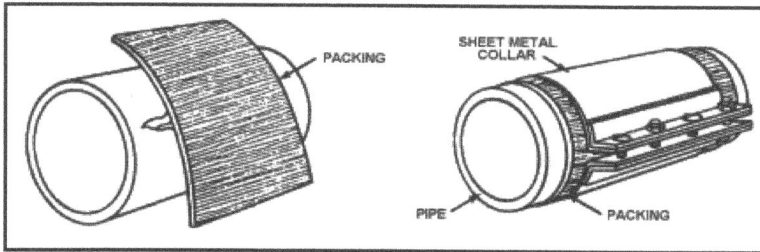

Figure 4-8. Jubilee Pipe Patches

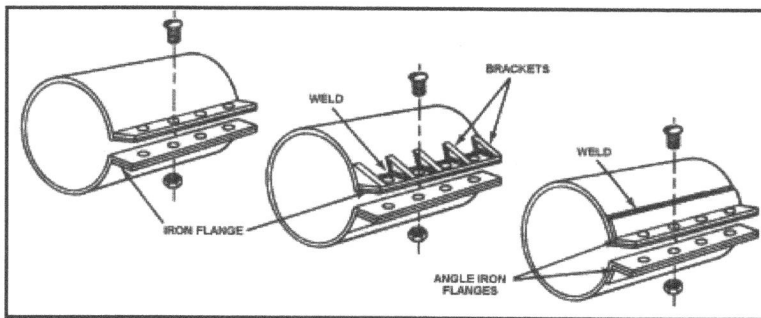

Figure 4-9. Three Types of Reinforced Metallic Clamps

4-32. **Metallic Pipe Repair Kit.** Most water, fuel, and gas lines can be repaired and restored to the system within 30 minutes if the contents of the emergency damage control metallic pipe repair kit (figure 4-9) are applied properly. In addition to repair or patching of piping, certain materials, which may be used to patch small cracks and ruptures in flat metal surfaces, are included in the kit. Materials in the kit may be obtained separately through appropriate supply channels whenever a need arises to replace them. You do not need to obtain another completely new kit.

ELECTRIC REPAIR KIT

4-33. The electrical tool kit contains, but is not limited to these items (figure 4-10):
- Chisel
- Pliers
- Hand Wire Stripper
- Hacksaws
- Ball peen hammer
- Screwdrivers
- Wrenches
- Tool belt
- Fuses
- Electrical tape

- Flashlight
- Line volt indicator
- Electrical workers gloves

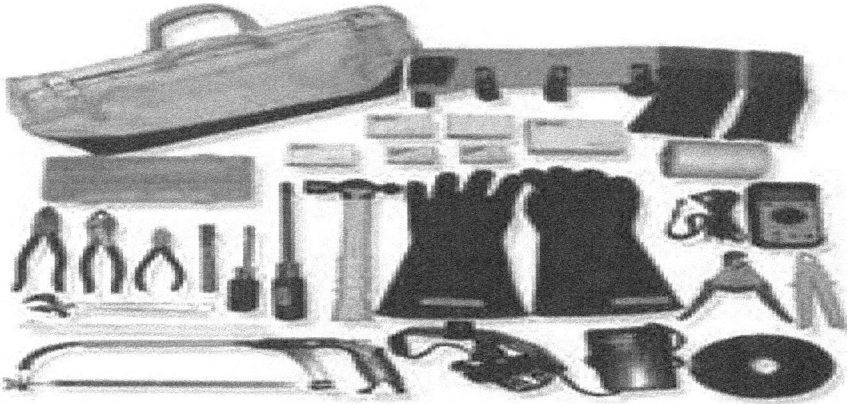

Figure 4-10. Electrical Tool Kit

4-34. The electrical tool kit is designed to repair faulty circuits providing that the hazardous area has been deemed safe enough to complete the work. After an unplanned event (i.e., an accident) all electrical systems should be thoroughly tested to ensure that further damage to equipment, machinery, or personnel will not arise.

4-35. When making electrical repairs, the following guidelines shall be adhered to by all Army watercraft personnel:

- No personnel at any time shall work on an electrical circuit or system alone.
- Know the potential hazard and use equipment properly.
- Prior to starting work ensure that all circuits of concern are deactivated and tagged out in two separate locations.
- Use a meter to ensure that circuit is de-energized at the source.
- Upon completion of work ensure that all tools are cleaned, in operational condition, and returned to proper storage area.

SHIP'S MAUL

4-36. The maul is a hand hammer with one flat striking surface and one wedge-shaped end (Figure 4-11). It is used to drive plugs and wedges into holes and cracks in piping and openings as well as for driving wedges into shoring structures. The maul is also used by the shoring watch for pounding shoring wedges and braces back into place should they become lose.

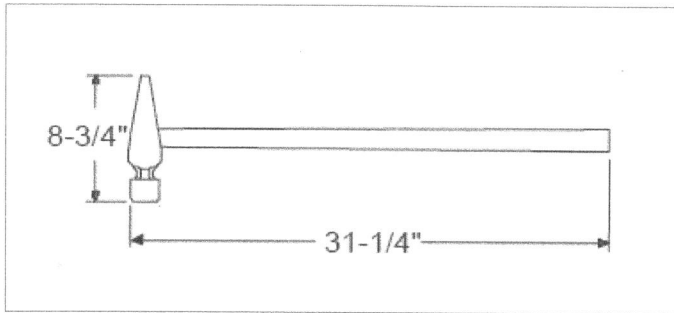

Figure 4-11. Ships Maul

SHORING

4-37. The purpose of shoring is to stop or reduce the inflow of water by patching, plugging, and supporting weakened or damaged areas of the vessel. Shoring may provide support either horizontally, upward or downward. Shoring horizontally or upward is called Bracing. To shore downward is to Tom.

Clearing the Decks

4-38. The first step in effective shoring is to clear the decks. Damage serious enough to produce a hole in the hull usually leaves wreckage scattered about the area. The damaged area should be cleared quickly to permit the damage control team to do a quick, adequate, and safe job.

4-39. Most loose wreckage can be removed by hand. At other times cutting and breaking are required. This requires the use of mauls, sledges, axes, heavy cold chisels, pinch bars, power drills, power chisels, and air hammers.

4-40. If fire accompanies the damage, burning bedding, stores, and supplies must be removed. A devil's claw (a homemade long-handled rake device) made of steel is handy for this purpose.

4-41. Shoring tools such as saws, 2-foot squares, hammers, and hatchets are stowed in the ship's damage control locker. Additional equipment may be devised, limited only by the ingenuity of the ship's crew.

Speed

4-42. Speed is critical in shoring. Seconds count, especially if there is a hole below the waterline. Each member of the damage control team must be able to think fast and improvise shoring with whatever material is available. Items such as life jackets, mattresses, pillows, and mess tables have proven to be satisfactory temporary shoring material.

Preparation

4-43. In order to make speedy repairs, preparation is vital. Only through regular drills can skills be developed that will enable each crew member to do a fast, effective shoring job under adverse conditions. A thorough training program should be established to train all crew members. Damage control lockers must be clean and orderly. All tools should be placed in secure mountings, yet readily detachable.

Shoring Principles

4-44. Observe the following basic principles when shoring damaged or weakened members of a ship's structure:

- **Spread the pressure.** Make full use of strength of members by anchoring shores (figure 4-12) against beams, stringers, frames, stiffeners, and stanchions. Place the legs of shoring against strong backs at angles from 45° to 90° (figure 4-13).

- **Plan shoring to hold the bulkhead as it is.** Do not attempt to force a warped, spring, or bulged bulkhead back into place (figure 4-14).

- **Secure all shoring.** Use nails and cleats to ensure that shoring will not work out of place. Use correct type of shoring, whether steel (figure (4-15) or wood (figure 4-16).

- **Inspect shoring periodically.** The motion of the ship can often produce new stresses that will cause even carefully placed shoring to work free. Inspect all shoring regularly, particularly when the ship is underway.

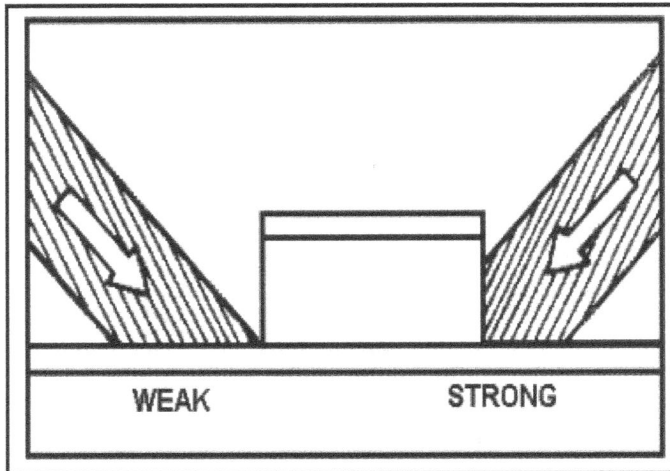

Figure 4-12. Anchoring Fit

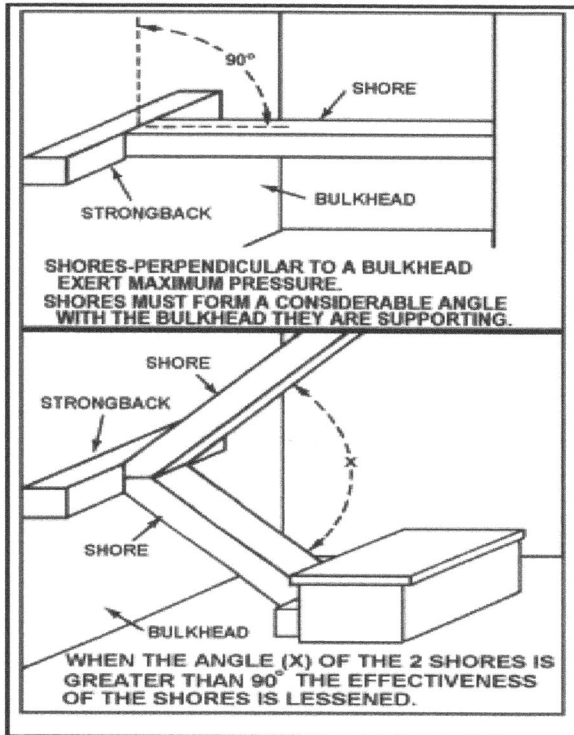

Figure 4-13. Correct Shoring Angles

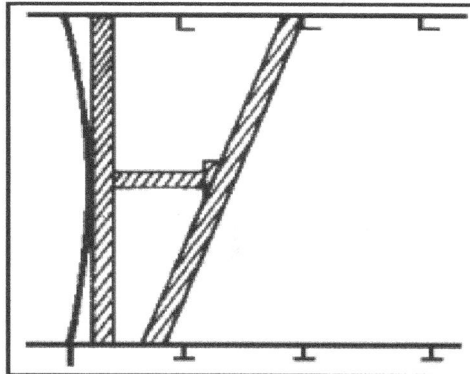

Figure 4-14. Shoring for Bulging Plate

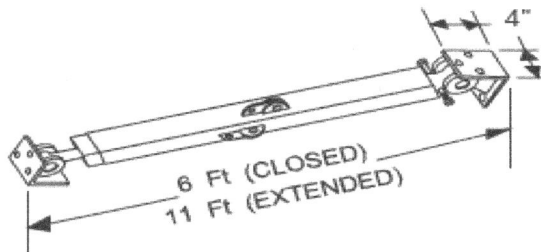

Figure 4-15. Steel Adjustable Shoring Batten

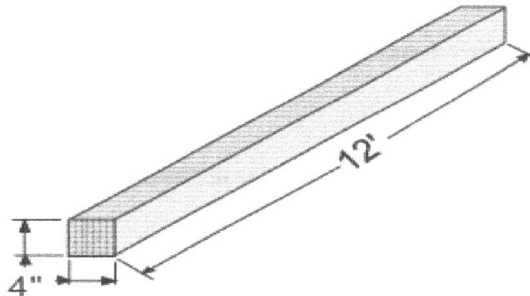

Figure 4-16. Wood Shoring

CONFINED SPACE ENTRY

4-45. This section outlines procedures established in the US Army Combat Readiness Center's Confined Spaces Program for the identification, preparation, testing, entry into and rescue from confined spaces within the Army watercraft fleet. On Army watercraft, when in port, Confined Space Entry procedures can only be initiated and performed by approved shore side personnel. Confined Space Entry procedures must be carried out properly by the vessel's crew when training for emergencies (i.e., fire or damage control) at sea, in port, and during actual events. Each ocean going Army watercraft has trained personnel and the equipment authorized to effectively evaluate the situation, plan and conduct the entry, perform any necessary personnel rescue, and completely clear the affected spaces to return those areas to inhabitable use.

```
┌─────────────────────────────────────────────────────────────┐
│                          WARNING                             │
│                                                              │
│   All spaces are Hot Work permit-required spaces for         │
│   welding operations.                                        │
│                                                              │
└─────────────────────────────────────────────────────────────┘
```

DEFINITIONS

4-46. The following terms are related to confined spaces:

- **Adjacent spaces.** Spaces bordering a subject space in all directions, including all points of contact, corners, diagonals, decks, tank tops, and bulkheads.
- **Attendant.** An individual stationed outside, a permit required confined space who monitors the authorized entrants and who performs all duties assigned in accordance with the unit's permit space program.
- **Authorized entrant.** An individual who has attended the required Confined Space Entry training either at the US Army Transportation School or National Fire Protection Association (NFPA) taught Competent Person course and is trained to enter a confined space.
- **Confined space.** A confined space is one that meets the following conditions:
  - Large enough that an individual can bodily enter and perform assigned work.
  - Has limited or restricted means for entry or exit (for example, tanks, vessels, vaults, pits, fuel cells, all rub railings, sealed ramp compartments, kort nozzles and rudders, stanchions and handrails, mast framework, and any sealed spaces onboard all watercraft).
  - Not designed for continuous occupancy.
- **Enclosed Space.** Any space, other than a confined space, which is enclosed by bulkheads and overheads. It includes cargo tanks, tanks quarters, and machinery and boiler spaces.
- **Emergency.** Any occurrence (including any failure of a hazard control or monitoring equipment) or event (internal or external) to the permit space that could endanger entrants.
- **Engulfment.** The surrounding and effective capture of a person by a liquid or finely divided (flowable) solid substance that can be aspirated to cause death by filling or plugging the respiratory system or that can exert enough force on the body to cause death by strangulation, constriction, or crushing.
- **Entry.** Any action resulting in any part of an individuals body breaking the plane of any opening of the confined space, and includes any work activities inside the confined space.
- **Entry permit.** The written or printed document that is provided by the Unit Safety Office to allow and control entry into a permit space and that contains the information specified in this manual.
- **Entry supervisor.** The person responsible for determining if acceptable entry conditions are present at a permit space where entry is planned, for authorizing entry and overseeing entry operations, and for terminating entry as required.
- **Hazardous atmosphere.** An atmosphere presenting a potential for death, disablement, injury, or acute illness from one or more of the following causes:
  - A flammable gas, vapor or mist in excess of 10% of its lower explosive limit (LEL).
  - An oxygen deficient atmosphere containing less than 19.5% oxygen by volume or an oxygen enriched atmosphere containing more than 22% oxygen by volume.
  - Airborne combustible dust at a concentration that meets or exceeds its LEL (airborne combustible dust which obscure vision at five feet or less).
  - Any other atmospheric condition that is immediately dangerous to life or health (IDLH).
- **Hotwork permit.** The written authorization to perform operations (such as welding, grinding, cutting, electrical drilling, etc) is capable of providing a source of ignition.
- **Immediately dangerous to life or health (IDLH).** Any atmosphere that poses an immediate threat to life or that is likely to result in acute or immediate severe health effects.

- **Lockout-tagout.** Placing locks and tags on energy isolating devices (e.g. breaker boxes, control switches, valves, etc.) to prevent the system from being re energized while work is still being performed by personnel.
- **Oxygen deficient atmosphere.** An atmosphere having an oxygen concentration of less than 19.5 percent by volume.
- **Oxygen enriched atmosphere.** An atmosphere that contains 22.0 percent or more oxygen by volume.
- **Permit-required confined space.** A space that meets the definition of a confined space and has one or more of these characteristics:
  - Contains or has a potential to contain a hazardous atmosphere.
  - Contains a material that has the potential for engulfing an entrant.
  - Has an internal configuration such that an entrant could be trapped or asphyxiated by inwardly converging walls or by a floor which slopes downward and tapers to a smaller cross section.
  - Contains any other recognized safety or health hazard.
- **Permit required confined space program.** The Command's overall program for controlling, and, where appropriate, for protecting personnel from permit space hazards and for regulating personnel entry into permit spaces.
- **Prohibited condition.** Any condition in a permit space that is not allowed by the permit during the period when entry is authorized.
- **Rescue teams.** The personnel designated to rescue personnel from permit spaces.
- **Retrieval system.** The equipment used for non-entry rescue of persons from permit spaces.
- **Testing.** The process by which the hazards, that may confront entrants of a permit space, are identified and evaluated. Testing includes specifying the tests that are to be performed in the permit space.

DUTIES AND RESPONSIBILITIES

The following identifies the duties and responsibilities of those involved with confined space entry:

4-47. **Commanders.** Commanders are responsible for the following:
- Establishing a confined space entry SOP within each organization conducting entry operations
- Ensuring recommended equipment to support confined space entry operations is purchased and maintained
- Ensuring personnel assigned confined space entry duties are adequately trained and certified
- Maintaining an inventory of all confined spaces within the organization
- Establishing risk approval procedures within the organization

4-48. **Entry Supervisor.** The entry supervisor is responsible for the following actions:
- Before Entry, the supervisor verifies that the permit is filled out completely all steps listed on it are taken, then signs the form.
- During Entry, the entry supervisor checks conditions to make sure they stay safe throughout the work.
- If conditions become unsafe, the permit is cancelled and everyone is ordered out of the space.
- The entry supervisor sees that any unauthorized people are removed.
- When the work is finished, the entry supervisor cancels the permit and concludes the operation.

4-49. **Entrant.** The entrant must:
- Know hazards that may be faced during entry.
- Be able to recognize signs or symptoms of hazard exposure and understand the consequences of such exposure.
- Use equipment properly.
- Maintain communication with the attendant.
- Alert the attendant to hazards discovered while in the space.
- Exit the space quickly when required.

4-50. **Attendant.** The attendant must:

- Know hazards that may be faced during entry.
- Be able to recognize signs or symptoms of hazard exposure.
- Maintain accurate entrant identification.
- Remain outside the space at all times.
- Maintain communication with the entrant and be able to communicate with the entry supervisor when needed.
- Monitor entry activities.
- Summon rescue services when needed.
- Prevent unauthorized entry.
- Perform non-entry rescue.
- Perform no conflicting duties.

SAFETY

4-51. In the past, over 60 percent of all fatalities in confined spaces were untrained rescuers. The primary cause of injury or death in confined spaces is asphyxiation. The second leading cause is fire. Implementation of the US Army Combat Readiness Center's Confined Spaces Program standard and application of these procedures will greatly reduce the potential for loss of life during entry into confined spaces.

CONFINED SPACE CLASSIFICATION

4-52. Each space must be evaluated on its own merit at the time of entry. Despite some leeway in the federal standards, you must treat all spaces as permit spaces prior to entry. This means that prior to entry, an entry supervisor will verify and document test results for oxygen content, flammability, and toxicity as well as evaluating other potential hazards. The entry supervisor may modify entry procedures based on this initial evaluation by classifying the space as non-permit. However, the space must clearly demonstrate no potential for developing a serious hazard during the work process. If a space is designated as non-permit, initial atmospheric test results must be documented and maintained on file for one year following the entry.

PREVENTING UNAUTHORIZED ENTRY

4-53. Each organization responsible for entering confined spaces must maintain an inventory of such spaces. The inventory must be organized so that the locations of referenced spaces are easily identified. All spaces that have the potential to contain atmospheric or other serious hazards must be marked using a "Danger Confined Space" sign and secured, if possible, to prevent unauthorized entry. All entries must be approved by a certified entry supervisor. During entry operations, an attendant will be positioned at the point of entry to ensure that only authorized entrants are allowed in the space. Attendants will summon the entry supervisor should unauthorized individuals interfere with safe operations. The entry supervisor will remove such individuals. To aid in preventing unauthorized entry, a safe zone must be identified around the point of entry using barricade tape or other means to warn individuals of a restricted area. Also, a "Danger Confined Space" sign must be posted at the point of entry.

CONFINED SPACE ENTRY EQUIPMENT

4-54. Organizations required to enter confined spaces will purchase and maintain a meter capable of measuring oxygen, flammability, and any toxic gases that can reasonably be expected to exist in the space atmosphere. The following Army watercrafts are authorized to acquire the necessary equipment: LSV, LT-800, and LCU Class 2000. DD Form 314, or other suitable records, must be maintained on all meters to document periodic, unit level calibration. Calibration must be done prior to each daily use, but will not exceed monthly if the meter is not in use. Organizations must ensure that calibration gas and replacement sensors are available when needed. Other equipment required for safe entry is identified on the entry permit and must be maintained and issued. Should such equipment not be available in the organization, entry may not proceed until it is obtained. Such equipment includes tripod with winch, lifeline and harness, non-sparking tools, lighting approved for hazardous atmospheres, ventilation blower, eye protection, hearing protection, gloves, and so on.

EVALUATING CONFINED SPACE HAZARDS

4-55. Use the following steps to evaluate confined space hazards:

- **Step 1-** The confined space must meet the following atmospheric criteria prior to entry:
- Percent of oxygen not below 19.5 percent or above 22 percent.
- Percent of LEL not above 10 percent.
- Parts per million of carbon monoxide not above 35 parts per million (ppm).
- Other atmospheric hazards not above the published permissible exposure limit (PEL). See the materiel safety data sheet (MSDS) for specific contaminants.

NOTE: Should there bean indication that other atmospheric hazards may exist but cannot be identified at the unit level, contact your supporting medical department activity (MEDDAC) for a consultation.

- **Step 2-** Visual inspection of the space prior to entry should identify other hazards that may exist. These may include noise, fall hazards, entrapment hazards, heat/cold, high pressure lines, inadequate lighting, and chemicals, piping carrying hazardous materials, moving machinery, electrical hazards, biohazards, radiation hazards, and so forth.

CONTROLLING CONFINED SPACE HAZARDS

4-56. The goal of each entry is to have optimum conditions. This means you should always strive to have 20.8 percent oxygen, 0 percent LEL, 0 ppm CO, 0 ppm of other hazardous gases, and fully control all other hazards. The entry supervisor is trained to establish these controls. Ventilation is the primary means of eliminating atmospheric hazards. The following are Basic ventilation standards:

- Meter readings must be taken prior to and after ventilating to evaluate the effectiveness.
- Normally, oxygen hazards will be controlled by blowing fresh air into the space and flammable/toxic hazards will be controlled by exhausting contaminated air from the space.
- If hazardous conditions are being created during the work process, ventilation may be needed continuously during the entry.
- Single point hazards such as welding and burning can best be controlled using local exhaust ventilation.
- When hazardous gases are heavier than air, exhaust low in the space and replace air high in the space. When hazardous gases are lighter than air, exhaust high in the space and replace air low in the space. Refer to the contaminants vapor density on the MSDS for determination.
- Should there be any doubt about the use of ventilation or its effectiveness, contact your supporting medial facility for a consultation.

4-57. Isolation of hazardous conditions is necessary before entry is allowed. Isolation is the process of ensuring that space remains free from release of energy or other hazards while the space is open for entry operations. The entry supervisor is responsible for evaluating hazards and the most effective means of isolation. Some controls include:

- Blanking and blinding.
- Removal of pipe sections.
- Double block and bleed.
- Lockout and/or tagout.
- Disconnecting mechanical linkages.
- Elimination of other hazards through cleaning, inserting, removal, guarding, reengineering, and soon should be the goal of the entry supervisor.

4-58. If hazards cannot be fully controlled, protective equipment must be used. Controls identified on the entry permit must remain in place during the work processor the entry must be terminated.

HOT WORK

4-59. Hot work includes any spark, flame, or extreme heat producing work such as welding, burning, brazing, grinding, cutting, chipping, use of tools that produce an electrical arc, and so on. The following procedures will be used to protect against the possibility of fire or explosion when performing hot work in or adjacent to confined spaces:

- A Marine Chemist, Army authorized person, or a Certified Industrial Hygienists will decide what work constitutes hot work prior to entry.
- If atmospheric testing shows or through evaluation of the work process a flammable environment is expected, both an entry permit and a hot work permit must be completed before work can begin.
- Hot work will not be accomplished in or near spaces containing more than 10 percent of LEL.

4-60. Organizations must make every attempt to engineer potentially flammable environments to 0 percent LEL before work commences or to fully control the potential hazard. This will be done by:

- Identifying all sources of flammable/combustible liquids, gases, and solids and using an acceptable means of isolating such sources from the space.
- Inserting spaces with a non-flammable inert gas if deemed appropriate by the entry supervisor. (Keep in mind that inserting creates an oxygen deficient atmosphere.)
- Cleaning and purging the space to remove flammable/combustible materials and residue.
- Covering combustible surfaces with a welding blanket or other suitable barrier.
- Using non-sparking tools and electrical appliances approved for the class and group of hazardous location expected
- Continuously monitoring LEL levels when the work process may produce a flammable/combustible atmosphere.
- Evaluating all adjacent spaces to ensure that there is no potential for igniting products in those areas.
- Having adequate fire extinguishing equipment on hand.
- Using a fire watch when necessary.

EMERGENCY RESPONSE

4-61. The following procedures will be used for an emergency response:

- The entry team is trained to identify symptoms of hazard exposure. The goal of each entry is to conduct self rescue in the case of any incident.
- The attendant or entrant, once an incident occurs, is obligated to clear the space immediately.
- If a meter alarm sounds with an entrant in the space, the space must be cleared immediately without first evaluating the reason for the alarm.
- Should an injury occur in the space and the entrant cannot conduct a self rescue, the attendant must initiate a rescue by contacting the installation fire department or other rescue service.
- Prior to the rescue team arriving, the attendant will notify the entry supervisor and may attempt a non-entry rescue; however, the attendant may not enter the space.

CONFINED SPACE ENTRY TRAINING

4-62. **Minimum standards for confined space entry training.** Each confined space entry team member must be trained. Initial Certification training is conducted at the US Army Transportation School or National Fire Protection Association (NFPA) taught Competent Person course. All training will be properly documented to include individual student identification.

PERMITS

4-63. Permits will be used according to the following.

- An entry permit will be used to document results of atmospheric tests and control safe entry.
- Only certified entry supervisors may approve entry permits.
- Entry permits will be approved for the minimum time necessary to complete operations.

- The original permit will be posted at the point of entry while work is in progress. – A copy of the permit will be maintained by the organization for three months following the entry.
- Only those entrants identified on the permit may enter the space.
- Hot work may require special precautions that are not identified on the entry permit. Should this occur, the permit will be used in conjunction with the entry permit. In such cases, the entry permit and the other permit will be posted at the point of entry.

COMMUNICATION

4-64. The entry supervisor will establish procedures for communication relative to each entry. Continuous communication between the entrant and attendant is required and emergency communication is required between the attendant and the rescue team and entry supervisor. Communication may be visual, verbal, signals, and so forth, but must be identified on the entry permit.

CONTRACTORS

4-65. Each organization must accomplish the following prior to contractor entries into confined spaces managed by this Command:

- Inform the contractor of permit program requirements.
- Apprise the contractor of safety precautions and procedures specific to the confined space being entered.
- Apprise the contractor of emergency response procedures on the installation.

CONCLUDING CONFINED SPACE ENTRIES

4-66. At the completion of work, the following will be done:

- Cancel the permit and file it for three months.
- Close the space and ensure that it is identified as a restricted area by having a "Danger Confined Space" sign posted at the point of entry.

HOT WORK IN CONFINED OR CLOSED SPACES

4-67. Hot work in the context of gas free engineering includes the following:

- Flame heating, welding, torch cutting, brazing, or carbon arc gouging. Chipping, electric drilling, grinding, electric wire brushing
- Any operation producing temperatures of 204.4 degrees C (400 degrees F)
- Any operation occurring in the presence of flammable materials or in a flammable atmosphere which requires the use or presence of an ignition source. Examples of such work include the following:
  - Spark-producing or static discharge.
  - Friction.
  - Open flames or embers.
  - Impact.
  - Non-explosion-proof equipment (such as lights, fixtures, or motors).

CAUTION

When open flame or heat producing work such as welding, cutting, or brazing is to be conducted, the worksite, regardless of the location, is to be inspected by the gas free engineer, safety NCO, fire department, or local approving authority as required by local SOP.

4-68. The provisions in this section apply to all hot work performed in confined or enclosed spaces, machinery rooms, bilges, and other locations proximate to flammable atmospheres (such as near fuel tank vents and sounding tubes). This section also applies to hot work performed on closed structures or containers such as pipes, drums, ducts, tubes, jacketed vessels, and similar items.

CLEANING AND VENTILATING FOR HOT WORK

4-69. Before hot work is begun in a confined or enclosed space, the space shall be tested, inspected, emptied of flammable cargo, cleaned, ventilated, and certified safe for hot work. Extraneous flammable or combustible materials such as scrap wood, paper, ropes, or rags shall be removed from the space. Combustible materials that cannot be removed shall be adequately protected. Ventilation ducting shall be made of noncombustible metal, of flexible construction, and free from hazardous combustible residues.

FIRE WATCH

4-70. When open flame or heat-producing work such as welding, cutting, or brazing is to be conducted, establish a trained fire watch at the worksite.

4-71. When hot work may transmit fire hazards to other spaces by overheating the connecting deck, overhead, or bulkheads, provide fire watches on both sides of the hot deck, overhead, or bulkheads.

4-72. Fire watch communications will enable the fire watch to report hazardous conditions on the opposite side of separating structures and provide a signal to stop hot work. Fire watches on both sides of the separating structure shall have, and know how to use, fire extinguishing equipment suitable to the exposure. Fire watches shall be equipped with personnel protective equipment as required for the operation being conducted (such as goggles, helmet, approved respiratory protective devices, and fire retardant clothing).

4-73. After completion of the hot work operation, fire watches shall remain on station for a minimum of 30 minutes; ensure the area is cool to the touch, and that no smoldering embers remain.

FIRE EXTINGUISHING EQUIPMENT

4-74. Fire extinguishing equipment shall be provided which is suitable for the nature and amount of flammable or combustibles present. Never use vaporizing liquid extinguishers in confined or enclosed spaces.

4-75. Use CO_2 extinguishers only after determining that the extinguisher is appropriate for the exposure. Also determine whether the displacement of oxygen by discharge of CO_2 into the space is likely to cause a hazard to personnel.

4-76. Water extinguishers or fire hoses equipped with Vari-nozzles, fog nozzles, or fog applicators are the most suitable fire extinguishing equipment for hot work in the presence of ordinary (class A) combustible material, flammable residues, coating, or insulation.

4-77. Evaluate fire extinguishing equipment for the following:
- Ability to suppress the fire
- Hazards that the extinguishing agent might create in the space.
- Capacity of the equipment compared to the fire potential. Fire hoses equipped with a Vari-nozzle, fog nozzle, applicator, or portable fire extinguisher are acceptable. The nature of the space or ship may restrict selection of fire equipment.

NOTE: Class A combustibles are those which leave embers and must therefore be cooled throughout the entire mass.

HOT WORK LOCATIONS

4-78. Prior to beginning hot work, an assessment of potential hazards must be made at each location. The following, although not all inclusive, provides guidance regarding what hazards to expect.
- **Boundary Spaces.** When hot work is to be performed on fuel tanks, associated vent spaces, or other spaces containing flammables (such as paint lockers and flammable liquid storerooms), the adjacent

spaces above, below, and on all sides (boundary spaces) shall first be inspected and tested, cleaned, and ventilated or inserted as appropriate, then certified "SAFE FOR HOT WORK".

- **Pipes, Tubes, and Coils.** Hollow connections to a space can present the same hazards as the space itself. Pipes, tubes, and coils or similar items which service, enter, or exit a confined or enclosed space shall be flushed, blown, purged, or otherwise cleaned and certified Safe for Hot Work before the performance of hot work on such items. If not so treated and certified, the certificate for the space shall be marked Not Safe for Hot Work. Valves to pipes, tubes, or similar items shall be dosed, the pipes blanked off, and tagged out, following the provisions of the Ship's Tagout Procedures, to prevent inadvertent discharge or backflow of materials into the space.

- **Hot Work on Closed Containers or Structures.** Prior to beginning hot work on hollow structures, drums, containers, jacketed vessels, or similar items, the items shall be cleaned, flushed, purged, made inert, filled with water, or otherwise made safe. The items shall be inspected, cleaned, tested, and certified before performing hot work. Items which are dosed and subject to pressure buildup from any application of heat shall be vented to relieve any pressure created by the hot work. The method of venting shall be selected to prevent ignition or explosion during the venting process.

- **Hot Work near Preservative Coatings.** Characteristics of a particular coating determine the procedures and precautions for hot work near that coating.

- **Flammable Coatings.** Flammable coating hot work requirements are as follows:

- Determine the flammability of coatings before starting hot work. Remove combustible coating from the hot work area to a distance sufficient to prevent ignition or outgassing (from temperature increase) at least 4 inches on all sides from the outer edge of the hot work.

- Never use flame or uncontrolled heat for stripping flammable coating. Test continuously for flammable atmospheres during hot work. Where significant outgassing is detected, stop hot work and further strip the coating. Start artificial cooling methods, such as wetting down, to prevent temperature increases in the un-stripped areas.

- Shield flammable coatings from slag or sparks in the area of the hot work. Wet down surrounding areas or cover with netted fire retardant cloth Ventilate area, if applicable.

- At a minimum, keep a 1-inch fire hose with a Vari-nozzle, fire nozzle, or fog applicator in the immediate vicinity, charged, and ready for instant use, except where prohibited by the nature of the space or ship.

- **Toxic Coatings.** Before hot work, strip any coating which becomes toxic when heated to at least 4 inches beyond the area that will be heated. Equip personnel with airline respirators or equivalent respiratory protection. Ventilate to remove toxic fumes or vapors from the space.

- **Hot Work near Damaged Surfaces.** Tank walls and coating deformities may carry toxicants and other hazards. Blisters, scales, and similar formations inside tanks that have held flammable materials may, even after cleaning and ventilating, hold flammable residues.

4-79. **Planning.** Consider the following when planning hot work:

- Determine whether any previous tank contents may have left hazardous residues.
- Assess the possibility of a surface flash which would involve the entire space.
- Clean Scales or Blisters. Consider the following when cleaning scales or blisters:
  - Remove scales or blisters which contain highly flammable residues (flashpoint 37.8 degrees C (100 degrees F) such as gasoline or jet propulsion (JP)-4 fuels from the entire space before hot work.
  - Clean away scales or blisters containing less flammable residues (flashpoint above 37.8 degrees C (100 degrees F) such as fuel oil or JP-5 fuel) to a distance of 4 inches on all sides from the outer edge of the hot work. In all cases, the area cleaned shall be sufficient to prevent out gassing from surrounding areas and to prevent ignition of residues.
  - Clean or protect areas below the hot work. Use screens, fire retardant cloth, or devices to capture sparks and slag.
  - Wet down areas around hot work to reduce the residue vaporization and to prevent small fires and flashes.
  - Assign fire watches with equipment to extinguish any resulting fire.

- Hot Work near pressurized systems. Before beginning hot work, depressurize nearby pressurized systems (such as hydraulics or Freon). Protect piping, fittings, valves, and other system components from contact with flames, arcs, slag, or sparks. Clean space and remove contaminants before hot work.

WARNING

When subjected to high temperatures, hydraulic fluid can decompose and produce highly toxic by-products. Noncombustible insulation such as fiberglass may have combustible backing or adhesive materials.

- Hot Work near Insulation. Conduct hot work carefully near combustible insulation. Some insulation materials may be ignited by welding, slag, or other short-duration exposure to ignition sources. Foam insulation materials are particularly likely to ignite and generate toxic combustion gases. The following are procedures for hot work near insulation:
  - Remove insulation from the area of hot work
  - Wet down non-removable insulation with water then cover the insulation with water soaked, fire-retardant cloth.
  - Station a fire watch with a charged 1-inch fire hose, ready for use in the immediate area.
  - Hot Work near Ammunition and Explosives. The following procedures are for hot work near ammunition or explosives:
- Remove ammunition and explosives from the area of hot work.
- Ventilate the area of the hot work.

HAZARDOUS BY-PRODUCTS

4-80. Welding, cutting, heating, or burning in the presence of certain materials (such as adulate fluids, Freon, chlorinated solvents, or Halon) can cause decomposition and produce hazardous by-products. Ensure that hot work is not conducted on or near such materials. Keep welding or cutting operations, which produce high levels of ultraviolet radiation, at least 200 feet from exposed chlorinated solvents.

GAS WELDING AND CUTTING OPERATIONS

4-81. The following shall be observed when performing gas welding or cutting operations:

- Transport, handle, and store compressed gas cylinders in accordance with the Code of Federal Regulations.
- Keep compressed gas cylinders or gas manifolds, used in welding and cutting operations, out of confined or enclosed spaces. Place them outside the space in open air, away from any fire, explosion, or emergency situation.
- Special care should be taken when working around electrical components.

COMPRESSED GAS CYLINDER STORAGE REQUIREMENTS

4-82. Compressed gas cylinders addressed here consist of any container which has a compressed gas, immaterial of size and content. Cylinders shall be stored at least 20 feet (6.1 m) from highly combustible materials such as oil or fuel. Cylinders should be stored in definitely assigned places away from stairs or gangways. Assigned storage places shall be located where cylinders will not be knocked over or damaged by passing or falling objects, or subject to tampering by unauthorized persons. Cylinders shall not be kept in unventilated enclosures such as lockers. Cylinder storage racks shall consist of both a raised area to set the cylinder into and a securing bracket near the top portion of the bottle. There should be no metal to metal contact between the cylinder and the storage rack. The cylinder valve head should have the screw type guard cap in place except when the cylinder is being used to provide its contents for its designed purpose. Installed or stowed compressed cylinders should be visually inspected monthly hydrostatically tested every 12 years or removed from service.

EMERGENCY LIGHTING

4-83. Emergency Lighting systems are both fixed and portable. Fixed systems comprise of both installed lighting system, marked with red letter "E", and semi-installed portable lanterns which are also wired directly into the vessel's emergency power system. Portable emergency lighting systems are stand alone, battery operated, lanterns mounted in specific areas throughout the vessel. Portable emergency lighting lanterns, Battle Lanterns, are marked with "EL" and the station number. Each semi-portable and portable emergency lighting station is labeled with the station number near the station location in addition to being marked on the lantern.

4-84. Emergency lighting and power systems test and inspection requirements are as follows:

a) Where fitted, it shall be the duty of the master to see that the emergency lighting and power systems are operated and inspected at least once in each week that the vessel is navigated to be assured that the system is in proper operating condition.

b) Internal combustion engine driven emergency generators shall be operated under load for at least 2 hours, at least once in each month that the vessel is navigated.

c) Storage batteries for emergency lighting and power systems shall be tested at least once in each 6-month period that the vessel is navigated to demonstrate the ability of the storage battery to supply the emergency loads for the specified period of time.

d) The date of the tests and the condition and performance of the apparatus shall be noted in the official logbook.

Chapter 5

Global Maritime Distress and Safety System

In 1979, the International Maritime Organization (IMO) recognized the need for an updated maritime communications system and helped create the International Maritime Satellite (INMARSAT) system employing geostationary satellites positioned above the Atlantic, Indian and Pacific oceans. Shortly thereafter, a polar orbiting satellite system was established to locate Emergency Position Indicating Radio Beacons (EPIRB). The IMO also decided to commence a general upgrade of the distress and safety system to be known as GMDSS. This system would provide rapid and automated distress reporting and improved telecommunications for the maritime community. In 1988, the IMO amended its Safety of Life At Sea (SOLAS) convention to complete this upgrade of the maritime safety communications procedures and equipment for GMDSS. GMDSS applies system automation techniques to the traditional maritime Medium Frequency (MF), High Frequency (HF) and Very High Frequency (VHF) bands, which previously required a continuous listening watch. GMDSS incorporates the INMARSAT and the EPIRB satellite systems to improve the reliability and effectiveness of the distress and safety system on a global basis. GMDSS also provides for the timely dissemination of maritime safety information, including navigational and meteorological warnings and weather forecasts.

On 1 February 1999, the voice watch keeping requirement on 2182 kHz for GMDSS equipped vessels ceased. The Coast Guard shore network maintains a voice guard on channel 16 VHF and 2182 kHz MF. These networks include the GMDSS Digital Selective Calling (DSC) on channel 70 VHF and 2187.5 kHz MF. While the Coast Guard plans to maintain the shore watch on channel VHF 16 for a number of years, there is no assurance that the 2182 kHz MF and HF voice watches will be continued. Existing Coast Guard MF and HF watches are being augmented with DSC to improve high seas telecommunications services to the maritime public.

As of 1 February 1999, all masters and mates require certification in accordance with the Coast Guard's Standards of Training Certification for Watchkeeping – 95 (STCW-95), which requires completion of a Coast Guard approved 70 hour course (fulfilled by the U.S. Army GMDSS course). The specific provisions requiring masters/mates to be trained/certified in GMDSS can be found at 43 CFR 10.205(n) for GMDSS operators, and 46 CFR 15.1103(e) and (f) for the "manning" requirements of two GMDSS operators on equipped vehicles.

OVERVIEW

DEFINITIONS OF SEA AREAS

SEA AREA A1

5-1.   An area within the radiotelephone coverage of at least one VHF coast station in which continuous digital selective calling (ch70) alerting and radiotelephony services are available, as defined by the International Maritime Organization. The United States presently has no A1 Sea Areas. Establishment of an A1 Sea Area in the U.S. is expected to depend upon approval and funding of the National Distress System Modernization Project.

Sea Area A2

5-2.   An area, excluding Sea Area A1, within the radiotelephone coverage of at least one MF coast station in which continuous DSC (2187.5 kHz) alerting and radiotelephony services are available, as defined by the International Maritime Organization (IMO). GMDSS-regulated ships traveling this area must carry a DSC-equipped MF radiotelephone in addition to equipment required for Sea Area A1. The United States presently has no declared A2 Sea Areas. The US Coast Guard has installed and is operating seven A2 Sea Area-capable coast stations, but those stations do not yet provide continuous coverage. Installation of DSC at most additional A2 Sea Area-capable coast stations is on hold, pending an upgrade to our 2 MHz (megahertz) infrastructure.

Sea Area A3

5-3.   An area, excluding sea areas A1 and A2, within the coverage of an INMARSAT geostationary satellite in which continuous alerting is available. Ships traveling this area must carry either an Inmarsat F77, B or C ship earth station, or a DSC-equipped HF radiotelephone/telex, in addition to equipment required for an A1 and A2 Area.

Sea Area A4

5-4.   The area outside that covered by areas A1, A2 and A3 is called Sea Area A4 Area. Ships traveling these Polar Regions must carry a DSC-equipped HF radiotelephone/telex, in addition to equipment required for areas A1 and A2.

DISTRESS ALERTING

5-5.   Distress alerting may be accomplished in three different ways: ship to shore, ship to ship and shore to ship. If terrestrial radio links, rather than satellite, are used, nearby ships will also hear the alert. The initial alert may be sent in a number of ways. The alert may be sent via INMARSAT-C, VHF/FM DSC radio, MF/HF DSC radio or EPIRB. All of these methods give the vessel's identity as well as its location. A DSC alert is the only type that can be picked up by another vessel. It is normally the responsibility of the Rescue Coordination Center (RCC) to respond with an acknowledgement. Vessels at sea should not normally acknowledge receipt of an initial distress alert.

DISTRESS RELAY

5-6.   Once an RCC has heard and acknowledged a distress, it may wish to alert other vessels in the area by means of a distress relay. The relay can be addressed to a precise geographic area so that vessels too far away to render help are not involved. Vessels can be alerted using INMARSAT-C, VHF/FM DSC radio, MF/HF DSC radio or Navigational TELEX (NAVTEX). Any vessel receiving a distress alert directly, or a distress relay, must contact the RCC to offer assistance. Vessels at sea should not normally send a distress relay themselves.

SEARCH AND RESCUE

5-7.   When the Search and Rescue (SAR) phase is entered, all communication is two-way to coordinate the activities of ships and aircraft using terrestrial and satellite communication links available. Specific frequencies are allocated for this purpose. Under all circumstances, a shore based RCC takes charge of the operation. The RCC may be located as much as a hemisphere away from the actual casualty. Vessels and aircraft close to the casualty will communicate between themselves using short range terrestrial communications (VHF or MF). Specially designated Search and Rescue (SAR) radio channels will be used. Precise location of the casualty will be aided by the use of a Search and Rescue Transponder (SART) or the 406 MHz section of a satellite Emergency Position Indicating Radio Beacon (EPIRB). Both of these items may be carried in the lifeboat. Portable VHF lifeboat radios are used by survivors to communicate with rescuers on channel 16, and one other channel, usually channel 6.

Recommended Publications

5-8.   Table 5-1 identifies the recommended publications to be maintained aboard all vessels equipped with GMDSS systems aboard.

NOTE: This list is not all inclusive, but should be filled as a minimum

Table 5-1. Recommended GMDSS Publications

NAME	QUANTITY
PUB 117 RADIO NAVIGATIONAL AIDS	1
MERCHANT SHIP SEARCH AND RESCUE MANUAL(MERSAR)	1
IMO SEARCH AND RESCUE MANUAL(IMOSAR MANUAL)	1
INTERNATIONAL TELECOMMUNICATIONS UNION(ITU):	
LIST OF CALL SIGNS AND NUMERICAL IDENTITIES	1
LIST OF COAST STATIONS	1
SHIP STATIONS(VOLUMES I, II, III)	1
CODE OF FEDERAL REGULATIONS TITLE 47 PART 80	1
COMSAT USERS GUIDE	1
IMO GMDSS OPERATORS GUIDANCE FOR MASTER OF SHIPS IN DISTRESS PLACARD IMO#969E	2

5-9.   The 406 MHz Category I EPIRB is used aboard Army Watercraft for electronic transmission of a data signal that will aid vessel/crew relocation in the event of capsizing, sinking, or abandon ship.

5-10. The 406 MHz Category I EPIRB is constructed of high impact resistant plastics are usually brightly colored. Generally, a four-position switch is incorporated that allows the unit to be armed, tested, disabled, or manually activated. A strobe light and antenna are also incorporated. The EPIRB is stored in a bracket that uses a hydrostatic release mechanism designed to allow automatic float free deployment and activation from the vessel when submerged to an approximate depth of 13 feet. Manual release and activation is also an option.

EMERGENCY POSITION INDICATING RADIOBEACON (EPIRB)

TYPES OF EPIRBS

Class A

5-11. 121.5/243 MHZ. Float-free, automatically-activating, detectable by aircraft and satellite. Coverage is limited. An alert from this device to a rescue coordination center may be delayed 4 - 6 or more hours.

Class B

5-12. 121.5/243 MHZ. Manually activated version of Class A. No longer recommended.

Class C

5-13. VHF ch15/16. Manually activated, operates on maritime channels only. Not detectable by satellite. These devices have been phased out by the FCC and are no longer recognized.

Class S

5-14. 121.5/243 MHZ. Similar to Class B, except it floats, or is an integral part of a survival craft. No longer recommended.

Category I

5-15. 406/121.5 MHZ. Float-free, automatically activated EPIRB. Detectable by satellite anywhere in the world. Recognized by GMDSS.

Category II

5-16. 406/121.5 MHZ. Similar to Category I, except is manually activated. Some models are also water activated.

121.5/243 MHz EPIRBs

5-17. These are the most common and least expensive type of Emergency Position Indicating Radio Beacon (EPIRB), designed to be detected by over flying commercial or military aircraft. Satellites were designed to detect these EPIRBs, but are limited for the following reasons:

- Satellite detection range is limited for these EPIRBs (satellites must be within line of sight of both the EPIRB and a ground terminal for detection to occur),
- Frequency congestion in the band used by these devices cause a high satellite false alert rate (99.8%); consequently, confirmation is required before search and rescue forces can be deployed,
- EPIRBs manufactured before October 1989 may have design or construction problems (e.g. some models will leak and cease operating when immersed in water), or may not be detectable by satellite. Such EPIRBs may no longer be sold,
- Because of location ambiguities and frequency congestion in this band, two or more satellite passes are necessary to determine if the signal is from an EPIRB and to determine the location of the EPIRB, delaying rescue by an average of 4 to 6 hours. In some cases, a rescue can be delayed as long as 12 hours.
- COSPAS-SARSAT is expected to cease detecting alerts on 121.5 MHz by 2008.
- 01 November 3, 2000, the National Oceanic and Atmospheric Administration (NOAA) announced that satellite processing 121.5/243 MHz emergency beacons will be terminated on February 1, 2009. Class A and B EPIRBs must be phased out by that date. The U.S. Coast Guard no longer recommends use of these EPIRBs.

406 MHz EPIRBs

5-18. The 406 MHz EPIRB was designed to operate with satellites. The signal frequency (406 MHz) has been designated internationally for use only for distress. Other communications and interference, such as on 121.5 MHz, is not allowed on this frequency. Its signal allows a satellite local user terminal to accurately locate the EPIRB (much more accurately -- 2 to 5 km vice 25 km -- than 121.5/243 MHz devices), and identify the vessel (the signal is encoded with the vessel's identity) anywhere in the world (there is no range limitation). These devices are detectable not only by COSPAS-SARSAT satellites which are polar orbiting, but also by geostationary GOES weather satellites. EPIRBs detected by the GEOSTAR system, consisting of GOES and other geostationary satellites, send rescue authorities an instant alert, but without location information unless the EPIRB is equipped with an integral GPS receiver. EPIRBs detected by COSPAS-SARSAT (e.g. TIROS N) satellites provide rescue authorities location of distress, but location and sometimes alerting may be delayed as much as an hour or two. These EPIRBs also include a 121.5 MHz homing signal, allowing aircraft and rescue craft to quickly find the vessel in distress. These are the only type of EPIRB which must be certified by Coast Guard approved independent laboratories before they can be sold in the United States.

5-19. A new type of 406 MHz EPIRB, having an integral GPS navigation receiver, became available in 1998. This EPIRB will send accurate location as well as identification information to rescue authorities immediately upon activation through both geostationary (GEOSAR) and polar orbiting satellites. These types of EPIRB are the most preferred.

5-20. 406 MHz emergency locating transmitters (ELTs) for aircraft are currently available. 406 MHz personnel locating beacons (PLBs) are available.

5-21. The Coast Guard recommends you purchase a 406 MHz EPIRB, preferably one with an integral GPS navigation receiver. A Cat I EPIRB should be purchased if it can be installed properly.

406 MHz GEOSAR System

5-22. The major advantage of the 406 MHz low earth orbit system is the provision of global Earth coverage using a limited number of polar-orbiting satellites. Coverage is not continuous, however, and it may take up to a couple of hours for an EPIRB alert to be received. To overcome this limitation, COSPAS-SARSAT has 406 MHz EPIRB repeaters aboard three geostationary satellites, plus one spare: GOES-W, at 135 deg W; GOES-E, at 75 deg W; INSAT-2A, at 74 deg E; and INSAT-2B (in-orbit spare), at 93.5 deg E. Ground stations capable of receiving 406 MHz. Except for areas between the United Kingdom and Norway, south of the east coast of Australia, and the area surrounding the Sea of Okhotsk near Russia, as well as polar areas, GEOSAR provides continuous global coverage of distress alerts from 406 MHz EPIRBs.

NOTE: The GEOSAR cannot detect 121.5 MHz alerts, nor can it route unregistered 406 MHz alerts to a rescue authority. GEOSAR cannot calculate the location of any alert it receives, unless the beacon has an integral GPS receiver.

THE COSPAS-SARSAT SYSTEM

5-23. COSPAS-SARSAT is an international satellite-based search and rescue system established by the U.S., Russia, Canada and France to locate emergency radio beacons transmitting on frequencies 121.5, 243 & 406 MHz.

COSPAS

5-24. Space System for Search of Distress Vessels (a Russian acronym)

SARSAT

5-25. Search and Rescue Satellite-Aided Tracking

NOTE: For more information on the COSPAS-SARSAT System go to http://www.cospas-sarsat.org/index.htm.

TESTING EPIRBS

5-26. The Coast Guard urges those owning EPIRBs to periodically examine them for water tightness, battery expiration date and signal presence. FCC rules allow Class A, B, and S EPIRBs to be turned on briefly (for three audio sweeps, or one second only) during the first five minutes of each hour. Signal presence can be detected by an FM radio tuned to 99.5 MHz, or an AM radio tuned to any vacant frequency and located close to an EPIRB. 406 MHz EPIRBs can be tested through its self-test function, which is an integral part of the device. 406 MHz EPIRBs can also be tested inside a container designed to prevent its reception by the satellite. Testing a 406 MHz EPIRB by allowing it to radiate outside such a container is illegal.

BATTERY REPLACEMENT

5-27. 406 MHz EPIRBs use a special type of lithium battery designed for long-term low-power consumption operation. Batteries must be replaced by the date indicated on the EPIRB label using the model specified by the manufacturer. It should be replaced by a dealer approved by the manufacturer. If the replacement battery is not the proper type, the EPIRB will not operate for the duration specified in a distress.

REGISTRATION OF 406 MHZ EPIRBS

5-28. Proper registration of your 406 MHz satellite emergency position-indicating radio beacon (EPIRB) is intended to save your life, and is mandated by Federal Communications Commission regulations. The Coast Guard is enforcing this FCC registration rule.

5-29. Your life may be saved as a result of registered emergency information. This information can be very helpful in confirming that a distress situation exists, and in arranging appropriate rescue efforts. Also, GOES, a geostationary National Oceanic & Atmospheric Administration (NOAA) weather satellite system can pick up and relay an EPIRB distress alert to the Coast Guard well before the international COSPAS-SARSAT satellite can provide location information. If the EPIRB is properly registered, the Coast Guard will be able to use the registration information to immediately begin action on the case. If the EPIRB is unregistered, a distress alert may take as much as two hours longer to reach the Coast Guard over the international satellite system. If an unregistered EPIRB transmission is abbreviated for any reason, the satellite will be unable to determine the EPIRBs location, and the Coast Guard will be unable to respond to the distress alert. Unregistered EPIRBs have needlessly cost the lives of several mariners since the satellite system became operational.

WHAT HAPPENS TO YOUR REGISTRATION FORM?

5-30. The registration sheet you fill out and send in is entered into the U.S. 406 Beacon Registration Database maintained by National Oceanic and Atmospheric Administration/ National Environmental Satellite, Data and Information Service (NOAA/NESDIS). If your EPIRB is activated, your registration information will be sent automatically to the appropriate USCG SAR Rescue Coordination Center (RCC) for response. One of the first things the RCC watch standers do is attempt to contact the owner/operator at the phone number listed in the database to determine if the vessel is underway (thus ruling out the possibility of a false alarm due to accidental activation or EPIRB malfunction), the intended route of the vessel if underway, the number of people on board, etc., from a family member. If there is no answer at this number, or no information, the other numbers listed in the database will be called to attempt to get the information described above needed to assist the RCC in responding appropriately to the EPIRB alert.

5-31. When RCC personnel contact the emergency phone numbers you provide, they will have all the information you have provided on the registration form. You should let these contacts know as much about your intended voyage as possible (i.e., intended route, stops, area you normally sail/fish/recreate, duration of trip, number of people going, etc.). The more information these contacts have, the better prepared our SAR personnel will be to react. The contacts can ask the RCC personnel contacting them to be kept informed of any developments, if they so desire.

REGISTRATION REGULATIONS

5-32. You may be fined for false activation of an unregistered EPIRB. The U.S. Coast Guard routinely refers cases involving the non-distress activation of an EPIRB (e.g., as a hoax, through gross negligence, carelessness or improper storage and handling) to the Federal Communications Commission. The FCC will prosecute cases based upon evidence provided by the Coast Guard, and will issue warning letters or notices of apparent liability for fine up to $10,000.

5-33. The Coast Guard has suspended forwarding non-distress activations of properly registered 406 MHz EPIRBs to the FCC, unless activation was due to hoax or gross negligence, since these search and rescue cases are less costly to prosecute.

5-34. If you purchase a new or a used 406 MHz EPIRB, you MUST register it with NOAA. If the EPIRB changes boat, unit, address, or primary contact phone number, you MUST re-register the EPIRB with NOAA. If the EPIRB is relocated from one vessel to another, make sure the new vessel re-registers the EPIRB, or the original vessel information may be called up by the Coast Guard if it later becomes activated.

5-35. An FCC ship station license is no longer required to purchase or carry an EPIRB.

HOW TO REGISTER

5-36. Download or request 406 MHz EPIRB registration forms from, http://www.sarsat.noaa.gov/beacon.html and mail or fax completed forms to the address listed below or call toll free at 1-888-212-SAVE (i.e. 1-888-212-7283) for further information or a copy of the registration form. From outside the U.S., call +1 (301) 457-5430 (fax: (301) 568-8649) for further information. Forms may be requested by phone or fax, or downloaded by computer (at above website). There is no charge for this service.

5-37. When filling out the registration form, fill out the EPIRB information as indicated in the Operator's Manual and on the identification labels. Do not enter individual soldier information in the Owner/Operator block. The vessel is an Army asset and Unit contact information should be entered. Unit telephone numbers listed should be the unit office phones, which do not change with a change of personnel. The master, mate or other GMDSS-trained individual should fill out the registration form. Enter information about the vessel and radio equipment onboard to include number of life rafts onboard. Radio call sign is the four (4) digit alphanumeric designation assigned to the vessel (ie. AAEF). The INMARSAT and MMSI will be found on GMDSS equipment. The Federal Registration Number is the Hull Number (ie. LSV-6, LCU 2023, TSV 1). Homeport is the port of the vessel's home station, such as 3rd Port, Fort Eustis, Virginia. The Primary 24-hour Emergency Contact is the Harbormaster of the vessel's home port. Include a fax number, if one is available. The Alternate 24-hour Emergency Contact is the Brigade Emergency Operations Center (EOC).

NOTE:

SARSAT BEACON REGISTRATION
E/SP3, RM 3320, FB-4
NOAA
5200 AUTH ROAD
SUITLAND MD 20746-4304

For more information see the NOAA SARSAT Homepage at http://www.sarsat.noaa.gov

SURVIVAL CRAFT TRANSCEIVER (SCT)

5-38. The 16/6 lifeboat radio is a portable two-way radiotelephone used for on-scene emergency communications between survival craft and rescue units. The radio is equipped with a five year lithium battery pack, which is operator replaceable. The radio will operate on channel 16 and one other channel designated for Search and Rescue, usually channel 6. The radio is FCC type accepted and GMDSS listed (FCC Part 80.1101) as a survival craft two-way VHF radiotelephone apparatus which complies with the 1988 GMDSS Safety of Life At Sea (SOLAS) amendments. The lifeboat radio should be tested semi-annually using a battery other than the assigned lifeboat radio battery. Three lifeboat radios are installed on the vessel.

Description

5-39. GMDSS Survival Craft Transceiver (SCT) is a portable, two-way VHF transceiver capable of radiotelephone communication. The equipment is capable of being utilized for on-scene communications between survival craft and rescue units. It may also be used between survival craft and the ship in distress, and for on-board communications when equipped with appropriate working channels.

Performance Standards

5-40. SCTs must meet IMO performance standards. The equipment is comprised of an integral transmitter, receiver (with push-to-talk switch), battery and antenna. A built-in microphone and speaker provides transmit and receive audio to the transceiver.

5-41. SCTs should be capable of withstanding drops onto a hard surface from a height of at least one (1) meter. Their watertight integrity must be maintained to a depth of one (1) meter for a period of at least five (5) minutes. SCTs have a lanyard or wrist strap which serves as an attachment to the user or his clothing.

EMISSION AND CHANNELS

5-42. SCTs must operate on 156.800 MHz (VHF channel 16) and at least one additional channel. This additional channel is often 156.300 MHz (VHF channel 06) because it is designated as an on-scene, Search and Rescue frequency. Most SCTs use phase modulation (G3E) in lieu of true Frequency Modulation (FM).

5-43. A number of commercially available SCTs operate on all VHF channels in the marine band. This permits the transceiver to meet the special demands of GMDSS and still meet daily operational requirements. All channels fitted are capable of single frequency voice communications.

BATTERY REQUIREMENTS

5-44. A number of SCTs use rechargeable NiCad batteries for day-to-day operations and a non-rechargeable lithium battery pack for maritime safety applications. The batteries are integrated into the transceiver. They must have sufficient capacity to ensure a minimum of 8 hours of operation at the SCTs highest RF power output.

5-45. Many of the newer model SCTs have batteries that may be replaced by the user. For batteries used for survival craft equipment, the month and year of its manufacture must be permanently marked on the battery. Also, the month and year upon which 50 percent of its useful life will expire must be permanently marked on both the battery and the outside of the transceiver.

5-46. Batteries must be replaced if 50% of their useful life has expired _or_ if the transmitter has been used in an actual emergency situation. Batteries must be replaced on or before the expiration date, *there is _no_ grace period*.

TRANSMITTER SPECIFICATIONS

5-47. The SCT must be capable of radiating a minimum Radio Frequency (RF) power of 250 mw (.25 watts) Many SCT transmitters are designed to produce 500 mw of power. If the transceiver is capable of producing RF power levels in excess of 1 watt, a power reduction switch must be provided to reduce drain on the battery. The power reduction switch must reduce RF transmitter power to 1 watt or less.

ANTENNA

5-48. SCT antennas must be vertically polarized. They should be tuned for maximum range and omni-directional in the horizontal plane. Some SCT manufacturers include a reflective tip on the antennas which may be used to attract the attention of rescue personnel.

CARRIAGE REQUIREMENTS

5-49. At least 3 two-way VHF Survival craft Transceivers must be provided on every passenger ship and cargo ships of 500 tons gross tonnage and upwards. At least 2 two-way VHF Survival Craft Transceivers must be provided on every cargo ship between 300-500 tons gross tonnage.

STOWAGE REQUIREMENTS

5-50. GMDSS SCTs must be stowed in such locations that they can be rapidly placed in any survival craft (other than life rafts required by the SOLAS convention). SCTs may be fitted as a fixed two-way VHF radiotelephone installation in survival craft. Such installations must adhere to the same performance standards of the portable SCT identified above.

TESTING REQUIREMENTS

5-51. Survival Craft Transceivers must be tested at intervals not to exceed 12 months.

SEARCH AND RESCUE TRANSPONDER (SART)

5-52. The Search and Rescue Transponder (SART) is a battery powered transponder used in an emergency by survivors of a sinking vessel. The SART must be mounted in the lifeboat one meter above the sea. The signal from the SART is detected by 9 GHz (3 cm) radar at a range of five to seven miles using the ship's radar. Aircraft radar can receive the SART signal flying at 3,000 ft at up to 40 nautical miles. Once activated, the SART will rebroadcast a response to a 3cm radar interrogation. At the same time, a line of 12 dots will appear on the search radar screen, radiating outwards from the position of the SART. Once the search vessel or aircraft has approached within one nautical mile of the SART, these dots widen to eventually form a series of concentric circles around the position of the SART. The SART has a built-in test capability and should be tested monthly. Two SARTs are installed on the vessel.

Description

5-53. GMDSS Search and Rescue Radar Transponders (SARTS) are portable devices capable of transmitting locating signals which indicate the location of a mobile unit in distress. The SART signal is picked up by the rescue unit's X -band, 9 GHz, 3 cm Navigational RADAR. The SART signal appears on the RADAR display as a series of equally spaced dots radiating outward on a line of bearing.

GENERAL

5-54. Search and Rescue Radar Transponders (SARTs) are the primary terrestrial means of providing locating signals in the GMDSS. They respond to interrogation by 3 cm, X-Band Radar from surface search vessels or aircraft.

ACTIVATION

5-55. SARTs are typically carried into survival craft and activated manually. SARTs are required to be capable of manual activation and deactivation. Provisions for automatic activation are also permitted. When activated from a ship in distress or survival craft, SARTs provide an audible and / or visual indication, to indicate proper operation, and to alert survivors whenever the RADAR has triggered the SART.

5-56. This indication should encourage survivors by letting them know rescue units are in the area. It also signals an appropriate time for survivors to attract the attention of rescue units via secondary alerting devices (e.g., SCT VHF Radio or pyrotechnics).

BATTERY CAPACITY

5-57. SARTs must have sufficient battery capacity to last for 96 hours in the standby mode followed by 8 hours in the transponder mode. The receiver has state of the art sensitivity to detect weak radar signals from search units. A SART antenna is typically horizontally polarized and Omni-directional. Some SARTs may have a desiccant cartridge or humidity indicator to detect if the unit's watertight integrity has been violated.

COMMUNICATIONS RANGE

5-58. Because SARTs rely upon direct wave propagation, their communications range is primarily determined by transmitter power output, receiver sensitivity and antenna height above ground. There is little affect a SART operator can have on the first two parameters, however, maintaining the SART as high as possible will optimize its effective range. Some SART manufacturers have included an extension mast to enable survivors to elevate the unit's antenna. The height of an installed SART antenna should be at least 1 meter above sea level.

PERFORMANCE STANDARDS

5-59. IMO performance standards for SART calls for a range of at least 5 miles when a SART is located 1 meter above sea level. This assumes the search vessel has a radar antenna 15 meters high in accordance with IMO requirements. An aircraft at an altitude of 3000 feet should be able to detect a SART at ranges up to 40 nautical miles (NM). The polar diagram of the SART antenna, its Omni-directional pattern and horizontal polarization are designed to optimize communications range, even in heavy swell conditions.

5-60. Most SARTs currently on the market are not designed to float free or be placed in the water, although IMO performance standards require SARTs to be capable of floating if not an integral part of a survival craft. SARTs are required to sustain a drop from 20 meters into the water without damage, and maintain watertight integrity to a depth of 10 meters for a period of at least 5 minutes.

CARRIAGE REQUIREMENTS

5-61. At least one radar transponder must be carried on every cargo ship of three hundred to five hundred gross tons and two radar transponders (one on each side) for every passenger ship and every cargo ship of 500 gross tons and upwards.

5-62. SARTs must be stowed in such locations that they can be rapidly placed in any survival craft.

NAVTEX

5-63. NAVTEX is an international automated direct printing service for the promulgation of navigational and meteorological warnings and urgent information to ships. It provides a low cost, simple means for the automatic reception of Marine Safety Information (MSI) by narrow band direct-printing telegraphy. NAVTEX is a component of the World Wide Navigational Warning Service (WWNWS) and is an essential element of the Global Maritime Distress and Safety System (GMDSS). Vessels regulated by the Safety of Life at Sea (SOLAS) Convention, as amended in 1988 (cargo vessels over 300 tons and passenger vessels, on international voyages), and operating in areas where NAVTEX service is available, have been required to carry NAVTEX receivers since 1 August 1993. The USCG discontinued broadcasts of safety information over MF Morse frequencies on that date. The USCG voice broadcasts (Ch. 22A), often of more inshore and harbor information, will remain unaffected by NAVTEX. A NAVTEX placard is intended to be laminated and either hung or posted near the NAVTEX receiver. Refer to Pub 117, Radio Navigational Aids for instructions.

NAVTEX FEATURES

5-64. NAVTEX messages are broadcast on a single frequency, 518 kHz, using the English language. Nominated stations within each NAVAREA transmit on a time-sharing basis to eliminate mutual interference. All necessary information is contained in each transmission. The power of each transmitter is regulated in order to avoid the possibility of interference between transmitters. A dedicated NAVTEX receiver has the ability to select messages to be printed according to a technical code (B1B2B3B4) which appears in the preamble of each message; and whether or not the particular message has already been printed.

5-65. By International agreement, certain essential classes of safety information such as navigational and meteorological warnings and search and rescue information are non-rejectable to ensure that ships using NAVTEX always receive the most vital information. NAVTEX coordinators exercise control of messages transmitted by each station according to the information contained in each message and the geographical coverage required. Therefore, the mariner may choose to accept messages, as appropriate, either from the single transmitter which serves the sea area around his position or from a number of transmitters.

MESSAGE PRIORITIES

5-66. Three message priorities are used to dictate the timing of the first broadcast of a new warning in the NAVTEX service. In descending order of urgency they are:
- VITAL–for immediate broadcast, subject to avoiding interference of ongoing transmissions
- IMPORTANT– for broadcast at the next available period when the frequency is unused;
- ROUTINE–for broadcast at the next scheduled transmission period.

5-67. Both VITAL and IMPORTANT warnings will normally need to be repeated, if still valid, at the next scheduled transmission period.

NAVTEX Message Selection

5-68. Every NAVTEX message is preceded by a four character header B(1)B(2)B(3)B(4). B(1) is an alpha character identifying the station, and B(2) is an alpha character used to identify the subject of the message. Receivers use these characters to reject messages from stations or concerning subjects of no interest to the user. B(3)B(4) are a two-digit number identifying individual messages, used by receivers to keep already received messages from being repeated. For example, a message preceded by the characters FE01 from a U.S. NAVTEX Station indicates that this is a weather forecast message from Boston MA.

TRANSMITTER IDENTIFICATION CHARACTER (B1)

5-69. The transmitter identification character B1 is a single unique letter which is allocated to each transmitter. It is used to identify the broadcasts which are to be accepted or rejected by the receiver. Two stations having the same B1 character must have a sufficient geographical separation so as to minimize interference with one another. NAVTEX transmissions have a designed range of about 400 nautical miles.

Table 5-2. NAVTEX Stations in the US B(1) Characters			
B(1) Character	*Station*	*Starting Time*	*Call Sign*
F	Cape Cod MA	0045Z	NMF
N	Chesapeake VA	0130	NMN
E	Savannah GA	0040	keyed by NMN
A	Miami FL	0000*	NMA
R	San Juan PR	0200	NMR
G	New Orleans LA	0300	NMG
C	Pt. Reyes CA	0000	NMC
Q	Cambria CA	0045	NMQ
W	Astoria OR	0130	NMW
J X	Kodiak AK**	0300 0340	NOJ
O	Honolulu HI	0040	NMO
V	Guam	0100	NRV

NOTE: Until a planned new automatic broadcast scheduler is installed, Miami's starting time of 0000 will be delayed approximately 5 minutes.

Kodiak also broadcasts safety information during time slots previously allocated to Adak.

SUBJECT INDICATOR CHARACTERS (B2)

5-70. Information in the NAVTEX broadcast is grouped by subject. The subject indicator character B2 is used by the receiver to identify the different classes of messages listed below. The indicator is also used to reject messages concerning certain optional subjects which are not required by the ship (e.g., LORAN-C messages might be rejected by a ship which is not fitted with a LORAN-C receiver). Receivers also use the B2 character to identify messages, which because of their importance, may not be rejected.

Table 5-3. NAVTEX Broadcast Subject Indicator Characters

A	Navigational warnings[1]
B[*]	Meteorological warnings[1]
C	Ice reports
D	Search & rescue information, and pirate warnings[1]
E	Meteorological forecasts
F[*]	Pilot service messages
G[*]	DECCA messages
H	LORAN messages
I	OMEGA messages (note OMEGA has been discontinued)
J	SATNAV messages (i.e. GPS or GLONASS)
K	Other electronic NAVAID messages
L	Navigational warnings (Additional to A)[2]
V to Y	Special services - allocation by IMO NAVTEX Panel
Z	No message on hand
[1] Cannot be rejected by receiver	
[2] Should not be rejected by receiver	
[*] Normally not used in the US	

NOTE: Since the National Weather Service normally includes meteorological warnings in forecast messages, meteorological warnings are broadcast using the subject indicator character E. U.S. Coast Guard District Broadcast Notices to Mariners affecting ships outside the line of demarcation, and inside the line of demarcation in areas where deep draft vessels operate, use the subject indicator character A. Two subject indicator characters for non-MSI messages in the United States were established 1 October 1995, but currently are not in use: V for Notice to Fisherman and W for Environmental messages.

MESSAGE NUMBERING (B3B4)

5-71. Each message within a subject group is assigned a two-digit serial number, B3B4, between 01 and 99. This number will not necessarily relate to series numbering in other radio navigational warning systems. On reaching 99, numbering should restart at 01 but avoid the use of message numbers still in force.

Technical Information

5-72. All NAVTEX broadcasts are made on 518 kHz, using narrow-band direct printing 7-unit forward error correcting (FEC or Mode B) transmission. This type of transmission is also used by Amateur Radio service (AMTOR). Broadcasts use 100 baud FSK modulation, with a frequency shift of 170 Hz. The center frequency of the audio spectrum applied to a single sideband transmitter is 1700 Hz. The receiver 6 dB bandwidth should be between 270-340 Hz.

5-73. Each character is transmitted twice. The first transmission (DX) of a specific character is followed by the transmission of four other characters, after which the retransmission (RX) of the first character takes place, allowing for time-diversity reception of 280 ms.

NOTE: For more information, see International Telecommunications Union (ITU) Recommendations M.540-2 and M.476-5, available from the ITU Radio communications Sector at http://www.itu.int/ITU-R/index.html.

PRACTICAL INSTRUCTIONS FOR THE USE OF A NAVTEX RECEIVER

5-74. The NAVTEX receiver is a Narrow Band Direct Printing (NBDP) device operating on the frequency 518 kHz (some equipment can also operate on 490 and 4209.5 kHz), and is a vital part of the Global Maritime Distress and Safety System (GMDSS).

5-75. It automatically receives Maritime Safety Information such as Radio Navigational Warnings, Storm/Gale Warnings, Meteorological Forecasts, Piracy Warnings, Distress Alerts, etc. (full details of the system can be found in IMO Publication IMO-951E - The NAVTEX Manual).

5-76. The information received is printed on the receiver's own paper recorder roll. Each message begins with a start of message function (ZCZC) followed by a space then four B characters. The first, (B1), identifies the station being received, the second, (B2), identifies the subject i.e. Navigational Warning, Met Forecasts, etc., and the third and fourth, (B3 + B4), form the consecutive number of the message from that station. This is followed by the text of the message and ends with an end of message function (NNNN).

5-77. The NAVTEX system broadcasts COASTAL WARNINGS that cover the area from the Fairway Buoy out to about 250 nautical miles from the transmitter; the transmissions from some transmitters can be received out to 400 nautical miles and even further in unusual propagation conditions.

5-78. The practical advice in paragraph 5-69 will help to ensure that you make the most efficient use of your NAVTEX receiver, guaranteeing the reception of Maritime Safety Information within the respective coverage areas of the NAVTEX stations being used.

NAVTEX RECEIVER CHECK-OFF LIST

5-79. For a NAVTEX receiver to function effectively, it is essential that the operator should have a sound knowledge of how to program and operate his particular receiver. This is not difficult provided the following practical steps are followed:

- Make sure that there are sufficient rolls of NAVTEX paper on board.
- Check that there is paper in the receiver.
- Turn the NAVTEX receiver on at least four hours before sailing, or better still, leaves it turned on permanently. This avoids the chance of losing vital information that could affect the vessel during its voyage.
- Make sure that the Equipment Operating Manual is available close to the equipment, paying particular attention to the fact that your equipment may be programmed differently from other makes and models.
- Using the Equipment Operating Manual, make a handy guide for programming, status and auto-testing procedures for your vessel's equipment, place it in a plastic cover and keep it with the equipment.
- Have available next to the equipment a plasticized copy of the NAVAREA's / METAREA's in which the vessel is likely to sail, showing the NAVTEX stations, their coverage ranges, their respective time schedules and B1 characters.
- Program your receiver to accept only those messages identified with the B1 character of the NAVTEX station which covers the area in which your vessel is currently sailing and the one covering the area into which you are about to sail. This will avoid the equipment printing information which has no relevance to your voyage and will avoid unnecessary waste of paper.
- Program your receiver to accept only those messages identified with the B2 characters (type of message) you wish to receive. It is recommended that most B2 characters (A to Z) be programmed, but you may exclude those for NAVAID equipment (Decca or Loran for example) with which your

vessel is NOT fitted. Be aware that the characters A, B and D MUST be included, as they are mandatory.

● Take extra care not to confuse the programming of B1 characters (station designators) with those of B2 characters (type of messages). It is very easy for an operator to believe that he/she is programming B1 characters when in fact they are programming B2 characters. After programming ALWAYS CHECK the program status to ensure that it is correct.

● If information is received incomplete/garbled, inform the relevant NAVTEX station, giving the time of reception (UTC) and your vessel's position. By so doing, not only will you obtain the information you require, but you will also help to improve the system. In the same way, any safety-critical occurrences observed during the voyage must be passed immediately to the nearest (or most convenient) Coast Radio Station and addressed to the relevant NAVAREA/METAREA or National Coordinator responsible for the area in which you are sailing.

INTERNATIONAL MARITIME SATELLITE ORGANIZATION (INMARSAT)

HISTORY

5-80. INMARSAT pioneered and developed global satellite communications. When INMARSAT began service in 1982 (with the Standard A system), its goal was to provide communications for commercial and distress and safety applications for ships at sea.

5-81. INMARSAT grew out of an initiative of the International Maritime Consultative Organization (IMCO), now the International Maritime Organization (IMO). At the time, mobile satellite communications was an unexplored technology and the industry an embryonic, untested one. So it was decided that INMARSAT should be a joint co-operative venture of governments, with their signatories (nominee organizations), in most cases the country's post and telecommunications providers, contributing the capital and bearing the high risk involved.

5-82. The purpose of INMARSAT is founded in the language of its original convention to make provision for the space segment necessary for improving maritime communications - thereby assisting and improving distress and safety of life, efficiency in management of ships and maritime public correspondence services." The convention also:

● Limits use of the system to "peaceful purposes"

● Makes the system available to ships of all nations

● Opens membership in the organization to all countries

5-83. Two decades after it was established, the Telecommunications Act of 1996 allowed INMARSAT to become the first intergovernmental "treaty" organization to privatize and become a limited company. INMARSAT headquarters is in London, and has regional offices in Beijing, Delhi and Dubai.

5-84. Today, INMARSAT partnership includes over 75 countries, 1,000 telecommunications companies providing services to over 140,000 plus Earth Station Terminal users worldwide.

5-85. The convention has since been revised to include aeronautical and land-mobile communications. While the maritime community constitutes the largest share of current users, an increasing number of land-mobile and aeronautical customers are changing the landscape of INMARSAT user demographics

INMARSAT - SYSTEM STRUCTURE

5-86. The INMARSAT system became operational on February 1, 1982. The system is designed to support a wide range of communications services to maritime, land-mobile and aeronautical users. The services provided by the INMARSAT system include: direct-dial telephone calls, telex messages, facsimile transmittals, electronic mail, and data transfer, emergency communications and automated position and status reporting.

5-87. Three basic segments comprise the INMARSAT system:

● THE SPACE SEGMENT

● THE GROUND SEGMENT

● EARTH STATIONS - SHIP/ MOBILE (SES) / (MES), COAST (CES) AND LAND (LES)

SPACE SEGMENT

5-88. The Space Segment is provided by INMARSAT, and consists of four communication satellites in a geostationary orbit positioned approximately 22,500 miles above the equator on their own meridian of longitude.

5-89. Each satellite has a coverage area (also known as a footprint), which is defined as the area on the earth's surface (sea or land) within which a mobile or fixed antenna can obtain line of sight communications with the satellite.

OCEAN REGIONS

5-90. Each satellite coverage area corresponds to an Ocean Region. The four Ocean Regions are:
- Atlantic Ocean Region-East (AOR-E)
- Atlantic Ocean Region-West (AOR-W)
- Indian Ocean Region (IOR)
- Pacific Ocean Region (POR)

SATELLITE POSITIONS

5-91. The INMARSAT satellite positions are:

Table 5-4. INMARSAT Satellite Positions

Ocean Region	AOR-E	AOR-W	IOR	POR
Position	15.5 W	54.0 W	64.5 E	178 E

- Based on a 5° elevation angle to the satellite, each ocean region can provide service between the latitudes of 70° North to 70° South. Each satellite provides coverage based on a maximum line of sight distance of 72° from Geographical Position (GP)

GROUND SEGMENT

5-92. The Ground Segment comprises a global network of:
- 38 Coast Earth Stations (CES)
- 1 Network Coordination Station (NCS) for each INMARSAT system within each Ocean Region,
- 1 Network Operations Center (NOC)

COAST EARTH STATION

5-93. Each CES/LES provides a link between the satellites and the national/international telecommunications networks. (i.e. Public Switch Telephone Network, PSTN)

5-94. A CES/LES operator is typically a large telecommunications company, which can provide a wide range of communications services to the Ship Earth Stations (SES) or Mobile Earth Stations (MES)

PRIORITY INDICATORS

5-95. CES transmit messages according to their priority of communications. There are four levels of priority within the INMARSAT system:
- Level 0 - Routine
- Level 1 - Safety
- Level 2 - Urgency
- Level 3 - Distress

NETWORK COORDINATION STATION

5-96. For each INMARSAT system a NCS is located within each Ocean Region, to monitor and control the communication traffic within its Ocean Region

5-97. Each NCS communicates with the CES/LES in its Ocean Region, and with other NCS to insure the proper and prompt transfer of information throughout the system is accomplished

NETWORK OPERATIONS CENTER

5-98. Overall responsibility for all INMARSAT communication systems and services

5-99. Monitors all CES/LES and NCS to insure that the transfer of information between all stations operates smoothly

5-100. Initiates remedial action when needed to prevent system overload

5-101. Located in the INMARSAT Headquarters, London, England

SHIP EARTH STATIONS

5-102. Ship Earth Stations (SES's) are communications terminals and their associated peripheral equipment installed aboard ship to enable two-way, multi-mode communications with shore side subscribers. There are separate SES's for each INMARSAT system A, B, and C.

5-103. The procedure used to initiate a vessel's SES into the INMARSAT system is commissioning through a Coast Earth Station (I.e., COMSAT).

5-104. INMARSAT A & B SES' can be purchased as either single channel or multi-channel models.

- Single Channel SES': A single channel SES is capable of using any one of its communication services at a time (I.e. phone, fax or telex), but not more than one service at a time.
- Multi Channel SES': Allows the user to operate more than one service at a time on different channels.

INMARSAT ABOVE DECK EQUIPMENT (ADE)

5-105. The INMARSAT B SES uses a highly directional, parabolic antenna (figure 5-1) approximately three (3) feet in diameter. INMARSAT B is a digital system that replaced the analog INMARSAT A, which is no longer recognized for GMDSS. The INMARSAT B antenna is controlled by an electronic control unit that accepts input from the ship's gyrocompass.

5-106. As the ship changes course, the input supplied to the control unit will compensate by keeping the antenna pointed at the satellite as the vessel comes to a new heading.

5-107. A stabilized antenna platform is maintained by a gimbaled assembly which compensates for ship rolling and pitching.

5-108. Finally, step tracking circuitry samples the satellite downlink signal periodically and makes minor adjustments to SES antenna position to optimize signal strength

Figure 5-1. INMARSAT B Above Deck Equipment

NOTE: SOME SYSTEMS WILL AUTOMATICALLY TRANSMIT DATA, SO EXTREME CAUTION MUST BE TAKEN WHEN NEAR AN INMARSAT A/B ANTENNA. RADIATION BURNS COULD RESULT.

INMARSAT B BELOW DECK EQUIPMENT (BDE)

5-109. The below deck equipment consists of:
- a power supply
- antenna control unit
- access control and signaling equipment
- satellite transmitter/receiver unit
- telex terminal
- telephones
- ancillary equipment (e.g., facsimile, PC, modem, etc.).

INMARSAT C ADE/BDE

5-110. INMARSAT C above-deck antenna is much smaller. One reason for this is that it is a store and forward system that passes data, but not voice signals. As a result, the usage cost is reduced and more vessels can participate in GMDSS. See figure 5-2 for examples of INMARSAT C above and below deck equipment.

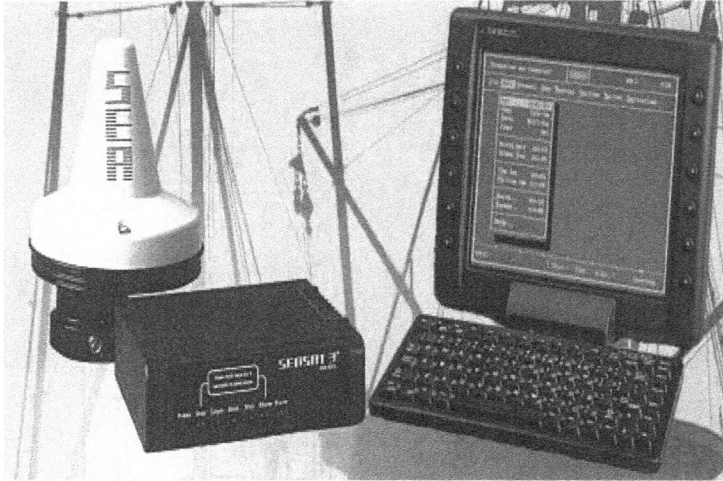

Figure 5-2. INMARSAT C Above/Below Deck Equipment

INMARSAT B

5-111. INMARSAT B is the most recent service introduced by INMARSAT. Digital INMARSAT B service has replaced analog INMARSAT A systems

5-112. INMARSAT B offers the same multi-mode functional capabilities of INMARSAT A, but because digital technology allows for more efficient use of the system, operational costs can be spread among a larger number of users.

5-113. INMARSAT B offers superior voice quality circuits at lower prices than INMARSAT A. More CES' supporting INMARSAT B service will be phased in during the coming years.

INMARSAT C

5-114. INMARSAT-C service was introduced in 1991. INMARSAT-C provides two-way store-and-forward messaging capabilities with small, low-cost SES terminals (i.e. laptop computers). See figure 5-2 for ADE/BDE.

5-115. INMARSAT-C messages encode information into Packet Oriented Protocol (POP) data for transmission over satellites at a rate of 600 bits per second. These messages are reformatted at the Coast Earth Station (CES) for delivery to facsimile, telex, data and e-mail systems ashore or to other SES'.

5-116. INMARSAT-C may also be used to interrogate ships at sea for their position, or for automatic data gathering at fixed or variable time intervals (Polling). INMARSAT C terminals are integrated with a Global Positioning System (GPS).

5-117. Forward Error Correction (FEC) and other techniques are used to insure the integrity of messages received at the CES. Messages are reassembled into their original form and sent to their destination via landline telecommunication networks.

5-118. The time frame for message delivery to a shore side subscriber is not precisely predictable, but most messages are delivered within three to eight minutes.

5-119. The global communications capability of the INMARSAT C system, combined with its Maritime Safety Information (MSI) broadcasting and distress alerting capabilities has resulted in the system being accepted by the IMO as meeting the requirements of the GMDSS. SOLAS now requires that INMARSAT C equipment have an integral satellite navigation receiver, or be externally connected to a satellite navigation receiver. That connection will ensure accurate location information to be sent to a rescue coordination center if a distress alert is ever transmitted.

5-120. INMARSAT C does not have voice capabilities.

LOSS OF COMMUNICATIONS

5-121. INMARSAT communications require direct wave connectivity between the satellite and the Ship Earth Station (SES). An obstruction, such as a mast, can cause disruption of the signal between the satellite and the SES antenna when the vessel is steering a certain course. This condition is sometimes referred to as "shadowing."

5-122. INMARSAT C systems are less likely to be effected by shadowing than A/B, but it could occur.

5-123. The most effective countermeasure to shadowing is to change course

5-124. Traveling beyond the effective radius of the satellite or use of a satellite whose signal is on a low elevation or below the horizon may also render INMARSAT communications impossible.

5-125. In marginal circuit conditions, it may be possible to initiate a reliable TELEX transmission but a voice communications may not be possible. Such conditions are indicated by low receiver signal strength on the SES terminal. Communications may become degraded or impossible when satellite elevation, as read on the INMARSAT A SES display becomes very small and is lowering.

5-126. Loss of satellite connectivity to an INMARSAT A/B unit may require the following data to be manually updated into the terminal:

- Satellites: azimuth & elevation of the satellite.
- Vessels: course, speed, & position of the vessel.

MF/HF RADIO

5-127. All ARMY vessels that operate in Sea Area A2, excluding Sea Area A1, within the radiotelephone coverage of at least one MF coast station in which continuous DSC (2187.5 kHz) alerting and radiotelephony services are available, as defined by the International Maritime Organization, must carry a DSC-equipped MF radiotelephone in addition to equipment required for Sea Area A1.

Table 5-5. US Coast Guard DSC-Equipped Shore Stations

Station	Type	Remote Site	MMSI
Chesapeake VA	MF/HF	CAMSLANT	003669995
COMMSTA Boston MA	MF/HF	Remote to CAMSLANT	003669991
COMMSTA Miami FL	MF/HF	Remote to CAMSLANT	003669997
COMMSTA Belle Chase LA	MF/HF	Remote to CAMSLANT	003669998
CAMSPAC Pt Reyes CA	MF/HF		003669990
COMMSTA Honolulu HI	MF/HF	Remote to CAMSPAC	003669993
COMMSTA Kodiak AK	MF/HF		003669899

VHF RADIO

5-128. Every Army vessel shall have on the bridge two operational VHF radios. A single VHF FM radio capable of scanning or sequential monitoring (often referred to as "dual watch" capability) will not meet the requirements for two radios. Channel 16 is the international distress and hailing frequency.

5-129. Use of the designated frequency (from 33 CFR, Part 26):

- No person may use the frequency designated by the Federal Communications Commission under section 8 of the Act, 33 U.S.C. 1207(a), to transmit any information other than information necessary for the safe navigation of vessels or necessary tests.

- Each person who is required to maintain a listening watch under Section 5 of the Act shall, when necessary, transmit and confirm, on the designated frequency, the intentions of his vessel and any other information necessary for the safe navigation of vessels.

- Nothing in these regulations may be construed as prohibiting the use of the designated frequency to communicate with shore stations to obtain or furnish information necessary for the safe navigation of vessels.

- On the navigable waters of the United States, channel 13 (156.65 MHz) is the designated frequency required to be monitored in accordance with §26.05(a) except that in the area prescribed in §26.03(e), channel 67 (156.375 MHz) is an additional frequency.

- On those navigable waters of the United States within a VTS area, the designated VTS frequency is the designated frequency required to be monitored in accordance with §26.05.

- Whenever radiotelephone capability is required by this Act, a vessel's radiotelephone equipment shall be maintained in effective operating condition. If the radiotelephone equipment carried aboard a vessel ceases to operate, the master shall exercise due diligence to restore it or cause it to be restored to effective operating condition at the earliest practicable time. The failure of a vessel's radiotelephone equipment shall not, in itself, constitute a violation of this Act, nor shall it obligate the master of any vessel to moor or anchor his vessel; however, the loss of radiotelephone capability shall be given consideration in the navigation of the vessel.

NOTE

AS STATED IN 47 CFR 80.148(B), A VHF WATCH ON CHANNEL 16 (156.800MHZ) IS NOT REQUIRED ON VESSELS SUBJECT TO THE VESSEL BRIDGE-TO-BRIDGE RADIOTELEPHONE ACT AND PARTICIPATING IN A VESSEL TRAFFIC SERVICE (VTS) SYSTEM WHEN THE WATCH IS MAINTAINED ON BOTH THE VESSEL BRIDGE-TO-BRIDGE FREQUENCY AND A DESIGNATED VTS FREQUENCY

DIGITAL SELECTIVE CALLING (DSC)

5-130. The IMO also introduced digital selective calling (DSC) on VHF, MF and HF maritime radios as part of the GMDSS system. DSC is primarily intended to initiate ship/ship, ship/shore, and shore/ship radiotelephone and MF/HF radio telex calls. DSC calls can also be made to individual ships or groups of ships. DSC distress alerts, which consist of a preformatted distress message, are used initiate emergency communications with ships and rescue coordination centers. DSC was intended to eliminate the need for persons on a ship's bridge or on shore to continuously guard radio receivers on voice radio channels, including VHF channel 16 (156.8 MHz) and 2182 kHz now used for distress, safety and calling. A listening watch aboard GMDSS-equipped ships on 2182 kHz ended on 1 February 1999.

5-131. IMO and ITU both require that the DSC-equipped VHF and MF/HF radios be externally connected to a satellite navigation receiver. That connection will ensure accurate location information is sent to a rescue coordination center if a distress alert is ever transmitted. FCC regulations actually require that ship's position be manually entered into the radio every four hours on ships required to carry GMDSS equipment, while that ship is underway. The Coast Guard believes VHF, MF and HF radiotelephone equipment carried on ships should include a DSC capability as a matter of safety. To achieve this, the FCC requires that all new VHF and MF/HF maritime radiotelephones type accepted after June 1999 to have at least a basic DSC capability.

SHIP RADIO AUTHORIZATION

5-132. Transmitter equipment will only be operated on those frequencies/channels authorized by the appropriate frequency management authority.

5-133. Operations will conform to Federal Communications Commission rules and regulations, part eight, rules governing stations on shipboard in the maritime service.

5-134. Ships operating on tactical nets using frequencies designated by competent authority will use tactical call signs designated by the same authority.

5-135. This ship radio authorization is authorized only for the periods indicated. If not renewed by the expiration date, the call sign will be assigned to another vessel.

5-136. When C-E equipment is permanently removed from the vessel, the vessel is sold, scrapped, or otherwise disposed of, or the vessel is transferred to another agency, the Department of the Army, NETCOM ESTA IPD office will be notified immediately, at (703) 325-8225.

CONTACT INFORMATION FOR RENEWAL OF RADIO AUTHORIZATION

Radio Frequency Authorization

5-137. This document is be posted so that any radio equipment and operators on the vessel can communicate on commercial, open channel, frequencies with approval by the Department of the Army. Only the Army Spectrum Management Office, Department of the Army, provides the legally required documentation. This document is issued only to the vessel by hull number and does not belong to any other entity. Radio Frequency Authorizations are valid for only three (3) years and must be renewed by requesting vessel/unit to:

SHIP RADIO AUTHORIZATION
Army Spectrum Management Office
2461 Eisenhower Ave, Suite 1204
Alexandria, VA 22331-2200
PH (703) 325-8225 or FAX 325-4138

This page intentionally left blank.

Chapter 6

GENERAL INFORMATION

SAFE TACTICAL WATERBORNE & CARGO OPERATIONS

GUIDANCE

6-1.   Army Regulation 385-10 contains additional specific guidance for waterborne safety.   Chapter 13 provides guidance for commanders of units conducting tactical water operations with vehicles.   Personnel operating vehicles embarking and disembarking a vessel require familiarization with techniques in vehicle swimming and fording procedures.   Chapter 14 provides additional safety requirements for cargo operations on all modes of transport, including hazardous material handling, port, supercargo, and escort standards.

BATTERY MAINTENANCE

BATTERY HANDLING

6-2.   There are different types of batteries such as lead-acid batteries, gel cells, and lead-calcium batteries. Most batteries contain sulfuric acid and lead. Because batteries contain chemicals, chemical reaction by-products, and an electrical current they can pose a hazard to soldiers if not handled properly. Soldiers that operate, maintain, and recharge batteries should use caution.

6-3.   Before working with batteries, Soldiers should have training in proper handling procedures. Personal protective equipment (PPE) should be worn at all times.   This includes:
- Chemical splash goggles
- Face shield
- Acid-resistant equipment such as gauntlet style gloves, an apron, and boots.
- In order to keep acid out of boots, do not tuck pant legs into boots.

6-4.   The sulfuric acid (electrolyte) in batteries is highly corrosive. Acid exposure can lead to skin irritation, eye damage, respiratory irritation, and tooth enamel erosion. Remember the following safety precautions:
- Never lean over a battery while boosting, testing or charging it.
- In marine environments, do not allow the battery solution to mix with salt water; it can produce hazardous chlorine gas.
- If acid splashes on the skin or eyes, immediately flood the area with cool running water for at least 15 minutes and seek medical attention immediately.
- Always practice good hygiene and wash your hands after handling a battery and before eating to prevent lead exposure. Signs of lead exposure include loss of appetite, diarrhea, constipation with cramping, difficulty sleeping, and fatigue.

6-5.   The chemical reaction by-products from a battery include oxygen and hydrogen gas. These can be explosive at high levels. Overcharging batteries can also create flammable gases. For this reason, it is very important to store and maintain batteries in a well-ventilated work area away from all ignition sources and incompatible materials. Cigarettes, flames or sparks could cause a battery to explode.

6-6.   Before working on a battery, disconnect the battery cables. To avoid sparking, always disconnect the negative battery cable first and reconnect it last. Be careful with flammable fluids when working on a battery-powered engine. The electrical voltage created by batteries can ignite flammable materials and cause severe burns. Soldiers have been injured and killed when loose or sparking battery connections ignited gasoline and solvent fumes during vehicle maintenance.

6-7.   Battery maintenance tools should be covered with several layers of electrical tape to avoid sparking. Place protective rubber boots on battery cable connections to prevent sparking on impact if a tool does accidentally hit a terminal. Clean the battery terminals with a plastic brush because wire brushes could create static and sparks. Always remove your personal jewelry before working on a battery. A short-circuit current can weld a ring or bracelet to metal and cause severe burns.

6-8.   Batteries can be very dense and heavy, so use proper lifting techniques to avoid back injuries. Battery casings can be brittle and break easily; they should be handled carefully to avoid an acid spill. Make sure that a battery is properly secured and upright in the vehicle or equipment. If a battery shows signs of damage to the terminals, case or cover, replace it with a new one. Finally, remember to dispose of old batteries properly. All parts are ordered by their national stock number (NSN).

EYE / FACE WASH STATIONS

6-9.   The Code of Federal Regulations states the following: "Where the eyes or body of any person may be exposed to injurious corrosive materials, suitable facilities for quick drenching or flushing of the eyes and body shall be provided within the work area for immediate emergency use." Chemical burns of the eyes need immediate first aid attention. Any delay in treatment will generally aggravate and intensify the injury.

6-10.   Initial treatment is actively flushing out the eyes with plenty of water. Irrigation should continue for a period of 15 to 30 minutes. This amount of time is usually adequate for the more serious chemicals. Though the initial flushing of the eyes or face is good, seek medical attention as soon as possible.

6-11.   The following units can be located at the eye/wash station:
*   **Plumbed Eye/Face Wash Units.** A plumbed eye/face wash unit is a permanently installed station that has a continuous supply of water. The supply line for plumbed units (figure 6-1) will provide an uninterruptible supply of water at approximately 30 psi. When installed, the actuation valve will be operated to determine that both eyes will be washed simultaneously at a velocity low enough not to cause injury to the user. The valve shall be designed so that the water flow remains on without requiring the use of the operator's hands. The valve shall be designed to remain activated until intentionally shut off. The valve should be simple to operate and shall go from "off" to "on" in one second or less. The valve shall be resistant to corrosion from potable water. The valve actuator will be large enough to be easily located and operated by the user. Plumbed eye/face wash units will be activated weekly to flush the line and to verify proper operation.
*   **Portable Eye/Face Wash Units.** Portable eye/face wash fountains (figure 6-2) generally are units which work on a gravity-fed system (normally holding 10 to 16 gallons of water). Self-contained units will be constructed of materials that will not corrode in the presence of the flushing fluid. There should be no sharp projections anywhere in the operating area of the unit. Nozzles shall be protected from airborne contaminants. Whatever means is used to afford such protection, its removal shall not require a separate motion by the operator when activating the unit. The unit will also be large enough to provide room to allow the eyelids to be held open with the hands while the eyes are in the stream of water. There is an anti-fungus additive for portable units to extend requirements for refilling according to the manufacturer recommendations. Water contained in system should be changed at regular interval so that any fungus or other possible contaminants are prevented from forming. Change out date should be posted on container so that the dated label is NOT easily removed.

NOTE: Every effort shall be made to install permanent eye/face fountains in all areas requiring an emergency eye/face wash capability.

6-12.   No portable eye/face wash units shall be permitted in areas where a chemical splash hazard exists and where there is a continuous source of clean water available. Portable eye/face wash fountains will be allowed in remote areas when no continuous flow of fresh water is available, when the installation of a fresh water system is not economically feasible, and when the hazard of chemical splash is minimal.

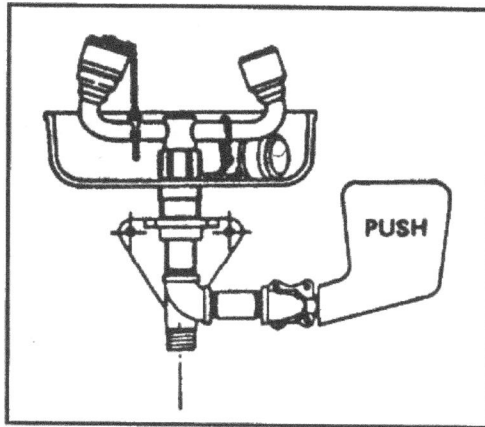

Figure 6-1. Plumbed eye / face wash unit (typical)

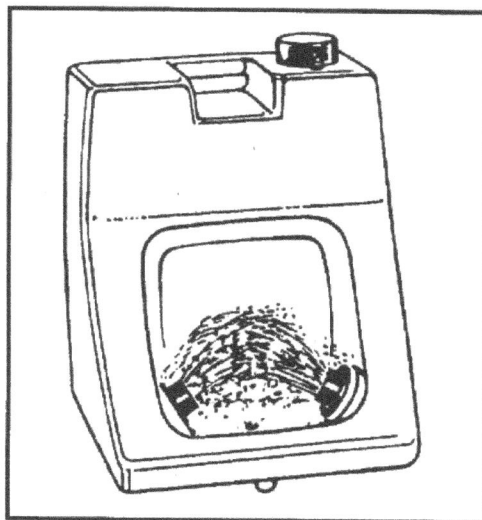

Figure 6-2. Portable eye / face wash unit

LOCATION

6-13. Eye/face wash units should be in accessible locations that require no more than 10 seconds to reach and should be within a travel distance no greater than 100 feet from the hazard. Specific installation instructions include that the unit be positioned about 45 inches from the floor. Each eye/face wash station shall be identified with a highly visible sign. The area around or behind, or both, the eye/face wash station will be painted a bright color and will be well lighted. If there is a specific working area that is used for only hazardous chemicals, then the wash station would be immediately adjacent to or within ten (10) feet.

TRAINING

6-14. All personnel who might be exposed to chemical splash will be instructed in the proper location and use of emergency eye/face wash stations.

POWER TOOL SAFETY

6-15. Power tools can be hazardous and have the potential for causing severe injuries when used or maintained improperly. Special attention toward hand and power tool safety is necessary in order to reduce or eliminate these hazards.

- Never carry a tool by the cord.
- Never yank the cord to disconnect it from the receptacle.
- Keep cords away from heat, oil, and sharp edges (including the cutting surface of a power saw or drill).
- Disconnect tools when not in use, before servicing, and when changing accessories such as blades, bits, etc.
- Avoid accidental starting. Do not hold fingers on the on/off switch button while carrying a plugged-in tool.
- Use gloves and appropriate safety footwear when using electric tools.
- Store electric tools in a dry place when not in use.
- Do not use electric tools in damp or wet locations unless they are approved for that purpose.
- Keep work areas well lighted when operating electric tools.
- Ensure that cords from electric tools do not present a tripping hazard.
- Remove all damaged portable electric tools from use and tag them: "Do Not Use."

COLOR CODING

6-16. Piping systems on Army watercraft are required to be color-coded. Valve handler and flow direction arrows are identified by colors. Additionally, for long piping runs, the piping should be marked either with the product name or color-coded banding. All fire main piping must be labeled "fire main" in addition to the above markings. Standardization of piping systems makes for easy identification by all crewmembers, who should be familiarized with the color codes and applications.

Table 6-1. Watercraft Color Codes

COLOR	APPLICATION
DARK GREEN	BILGE SYSTEM, SEA WATER SYSTEM
PURPLE	REFRIGERANT SYSTEM
DARK GRAY	HIGH PRESSURE AIR SYSTEM
TAN	LOW PRESSURE AIR SYSTEM
YELLOW	DIESEL FUEL SYSTEMS CONTAINERS- PAINT ENTIRE NAPTHA CONTAINERS- CONTAINER
ORANGE	LUBRICATING OIL HYDRAULIC OIL
RED	FIRE EXTINGUISHING SYSTEM
BLACK	STEAM AND HOT WATER HEATING SYSTEM
DARK BLUE	NON-POTABLE FRESH WATER SYSTEM
LIGHT BLUE	POTABLE WATER SYSTEM
GOLD	SEWAGE
YELLOW	OIL WATER SEPERATER SUCTION

PERSONAL PROTECTIVE EQUIPMENT (PPE)

EYE AND FACE PROTECTION

6-17. Soldiers must use appropriate eye or face protection when exposed to eye or face hazards from flying particles, molten metal, acids or caustic liquids, or other liquid chemicals, chemical gases or vapors, or potentially hazardous light radiation sources.

6-18. Face shields are used in operations when the entire face needs protection and to protect the eyes and face against flying particles, metal sparks, and chemical or biological splash hazards. Face shields must be used in combination with goggles when there is a potentially significant chemical splash hazard or when there is a potentially severe exposure to flying fragments or objects, hot sparks from furnace operations, potential splash from molten metal, or extreme temperatures.

6-19. Standard safety glasses are designed to protect against flying particles. Soldiers must use safety eyewear with side protection when there is a hazard from flying objects. Detachable side protectors (e.g. clip-on or slide-on shields) are acceptable if they meet the ANSI requirements.

6-20. Goggles offer the best all-around impact protection of all eyewear types because they form a positive seal around the eye area. Welders' goggles provide protection from sparking, scaling, or splashing metals and harmful light rays. Welding shields must be provided to protect soldiers' eyes and face from infrared or radiant light burns, flying sparks, metal spatter and slag chips encountered during welding, brazing, soldering, resistance welding, bare or shielded electrical arc welding, and oxyacetylene work. Lenses are impact resistant and are available in graduated shades of filtration.

6-21. Soldiers must use equipment with filter lenses that have a shade number appropriate for the work being performed for protection from injurious light radiation. Tinted and shaded lenses are not filter lenses unless they are marked or identified as such.

6-22. Soldiers who wear prescription lenses, while engaged in operations that involves eye hazards, must either:

- Wear eye protection that incorporates the prescription in its design, or,
- Wear eye protection that can be worn over the prescription lenses without disturbing the proper position of the prescription lenses or the protective lenses.

MAINTENANCE OF PROTECTIVE EYEWEAR

6-23. Safety glasses and other eye and face protection should be stored carefully so they won't be scratched or damaged. In general, do not store this equipment where it would be exposed to high heat or sunlight.

6-24. Inspect eye and face protection prior to use. If the equipment is damaged or broken, do not use it because it may not be able to fully resist impact.

6-25. Pitted lenses, like dirty lenses, make it more difficult for an employee to see and should be replaced. Lenses that are pitted or deeply scratched are more prone to break under impact and should be replaced.

6-26. Clean eye and face protection according to the manufacturer's instructions. If the manufacturer's instructions are not available, clean with a mild soap and water solution by soaking in the soap solution for ten minutes. Rinse thoroughly and allow to air dry.

GLOVES

6-27. Gloves are worn when the possibly of injury to either fingers or hands exist. Examples of possible injury occurring are during line handling, painting, moving or lifting heavy objects, securing or unsecuring cargo, rigging tows using wire cable, chain, or synthetic lines.

EQUIPMENT GUARDS

6-28. Equipment guards come in many shapes and sizes, each depending on the usage. Machinery guards are used to prevent injuries from any moving equipment. In areas that may contain any type of fumes, explosion-proof light globes with wire cages are found.

6-29. **Machine Guard Types**:
- Fixed
- Interlocked
- Adjustable
- Self-Adjusting

6-30. **Fixed Guards.** Fixed guards are:
- A permanent part of the machine
- Not dependent on any other part to perform the function
- Usually made of sheet metal, screen, bars or other material which will withstand the anticipated impact
- Generally considered the preferred type of guard.
- Simple and durable

6-31. **Interlocked guards**:
- Are usually connected to a mechanism that will cut off the power automatically.
- May be used on electrical, mechanical or hydraulic systems.
- Should rely on a manual reset system.
- Are adjustable.

6-32. **Adjustable Guards.** Adjustable Guards are very flexible to accommodate various types of stock. They are self-adjusting, but may also be adjusted manually.

6-33. **Self-Adjusting.** The opening of the self-adjusting is determined by the movement of the stock through the guard. This type of guard does not always provide maximum protection

6-34. **Explosion Proof Guards.** Explosion proof guards are made of heavy clear glass or plastic lens and screw over a light bulb into the light fixture. Depending on the type, many explosion proof globes also have wire cages over the clear lens. These cages are also screwed over the lens. The globe is designed to contain any spark which may be caused by light bulb should it be broken. The cages are designed to prevent impact to both the globe and bulb.

FALL PROTECTION

FALL PROTECTION SYSTEMS

6-35. Fall protection systems can consist of devices that arrest a free fall or devices that restrain a soldier in position to prevent a fall from occurring (figures 6-3 to 6-6). A **fall arrest system** is employed when a soldiers is at risk of falling from an elevated position. A **positioning system** restrains the elevated soldiers, preventing him from getting into a hazardous position where a fall could occur, and also allows hands-free work. Both systems have three components: harnesses or belts, connection devices and tie-off points.

HARNESSES AND BELTS

6-36. **Full-body harnesses.** This type of harness wraps around the waist, shoulders and legs. A D-ring located in the center of the back provides a connecting point for lanyards or other fall arrest connection devices. In the event of a fall, a full-body harness distributes the force of the impact throughout the trunk of the body—not just in the abdominal area. This allows the pelvis and shoulders to help absorb the shock, reducing the impact to the abdominal area.

6-37. Maximum force arrest on a full-body harness, which is used for the most severe free fall hazards, is 1800 pounds. Full-body harnesses come with optional side, front and shoulder D-rings. The side and front D-rings are connection points used for work positioning, and the shoulder D-rings are for retrieval from confined spaces.

6-38. Three factors determine the arresting force from a fall: lanyard material type, free fall distance and the weight of the soldiers. The use of a shock-absorbing lanyard or a higher tie-off point will reduce the impact force.

6-39. **Belts.** Belts are used in positioning system applications. These belts have two side D-rings, and are used only for restraining a soldiers in position. This type of belt is not used for any vertical free fall protection.

CONNECTION DEVICES

6-40. Connection devices attach the belt or harness to the final tie-off point. This can be one device, such as a lanyard, or a combination of devices, such as lanyards, lifelines, work lines, rope grabs, tie-off straps and safety hooks.

Lanyards

6-41. Lanyards are used both to restrain soldiers in position, and to arrest falls. When using a lanyard as a restraining device, the length is kept as short as possible, as a restraining lanyard should not allow a Soldier to fall more than two feet. Restraining lanyards are available in a variety of materials, including steel cables, rebar chain assemblies and nylon rope. Fall protection lanyards can be made of steel, nylon rope, or nylon or Dacron webbing. Fall protection lanyards may also have a shock-absorbing feature built in, thus reducing the potential fall arrest force. Remember that maximum arrest force is 900 pounds for belts, or 1800 pounds for full-body harnesses. With a belt, the use of a shock-absorbing lanyard is recommended because it limits the arresting force from a six-foot drop to 830 pounds. If a shock-absorbing lanyard is not used, the tie-off point must be high enough to limit the arrest force to less than the 900-pound limit. The height of this tie-off point will vary, depending on the lanyard material and the weight of the person involved. A lanyard used for a fall is limited to allow a maximum six-foot free fall. For this reason, most lanyards are a maximum of six feet long. However, if a higher tie-off point is used, the lanyard can be longer if the free fall distance does not exceed 6 feet.

LIFELINES/SAFETY HOOKS

6-42. **Lifelines.** Lifelines add versatility to the fall arrest system. When used in conjunction with **rope grabs**, a lifeline allows the soldiers to move along the length of the line rather than having to disconnect and find a new tie-off point. The rope grab is engineered to arrest a fall instantly. A rope grab and lifeline system is a passive form of protection, allowing the user to move as long as tension is slack on the lifeline. If a fall occurs, the tension on the rope grab triggers the internal mechanism to arrest the fall. **Retractable lifelines** automatically retract any slack line between the soldiers and the tie-off point. While this type of line doesn't require a rope grab, it must be kept directly above the soldiers to eliminate any potential swing hazard if the Soldier falls.

6-43. A **cross-arm strap** is used at a tie-off point with a large diameter, such as an I-beam, to which a lanyard or lifeline cannot directly attach. Using a cross-arm strap ensures the lanyard or lifeline doesn't become abraded from wrapping around the I-beam. A safety hook works in the same situations. It is used for tie-off points with a diameter of one to five inches, and then the lanyard is attached to the safety hook.

6-44. **Tie-off Points.** A **tie-off point** is where the lanyard or lifeline is attached to a structural support. This support must have a 5000-pound capacity for each soldiers tying off. Soldiers must always tie off at or above the D-ring point of the belt or harness. This ensures that the free fall is minimized, and that the lanyard doesn't interfere with personal movement. Soldiers must also tie off in a manner that ensures no lower level will be struck during a fall. To do this, add the height of the soldiers, the lanyard length, and an elongation factor of 3.5 feet. Using this formula, a six-foot tall soldier requires a tie-off point at least 15.5 feet above the next lower level.

6-45. **Other Devices.** For confined space applications, a **tripod and winch system** is used as both the tie-off point and connection device. It is used in conjunction with a full-body harness to lower and raise soldiers into tanks or manholes. Make sure that the tripod system you choose is designed for your application. Never **use a material-handling device for personnel** unless it is specifically designed to do so.

6-46. **Ladder systems** are lifelines attached directly to a ladder. The systems consist of a cable or channel, with a grabbing device attached for a connection point.

INSPECTION AND MAINTENANCE

6-47. Regulations require that all fall arrest equipment be inspected prior to its use. This includes looking for frays or broken strands in lanyards, belts and lifelines, and oxidation or distortion of any metal connection devices. To properly maintain the devices, periodic cleaning is necessary. Clean all surfaces with a mild detergent soap, and always let the equipment air dry away from excess heat. Follow the manufacturer's instructions for cleaning and maintenance.

NOTE: ANY EQUIPMENT EXPOSED TO A FALL MUST BE TAKEN OUT OF SERVICE AND NOT USED AGAIN FOR FALL PROTECTION.

Components of Fall Protection Systems:

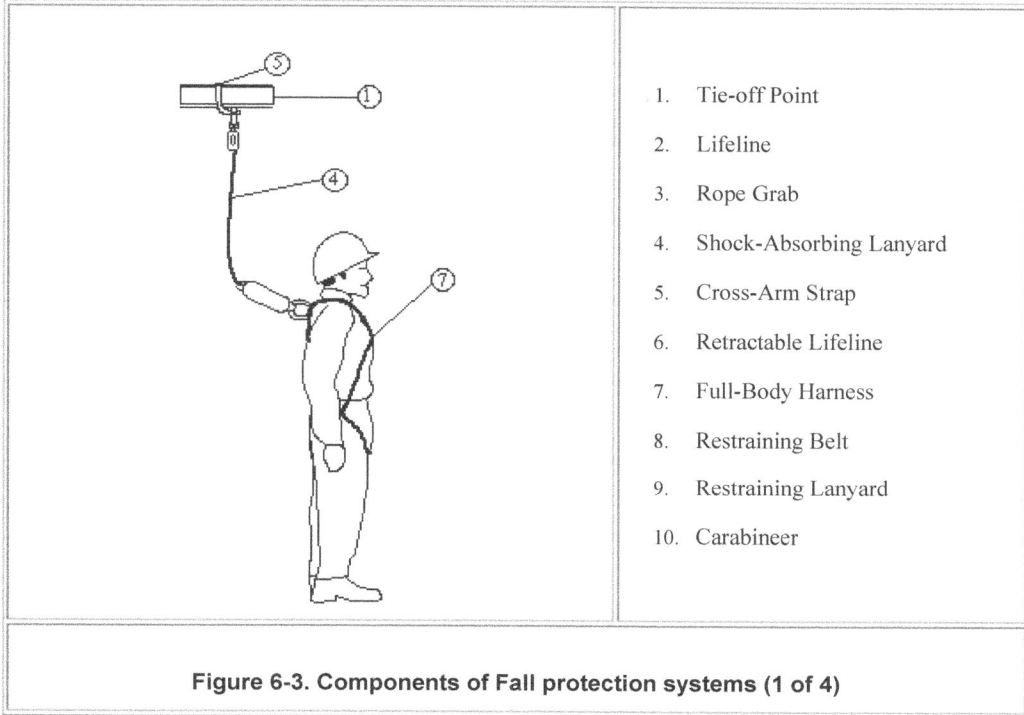

1. Tie-off Point
2. Lifeline
3. Rope Grab
4. Shock-Absorbing Lanyard
5. Cross-Arm Strap
6. Retractable Lifeline
7. Full-Body Harness
8. Restraining Belt
9. Restraining Lanyard
10. Carabineer

Figure 6-3. Components of Fall protection systems (1 of 4)

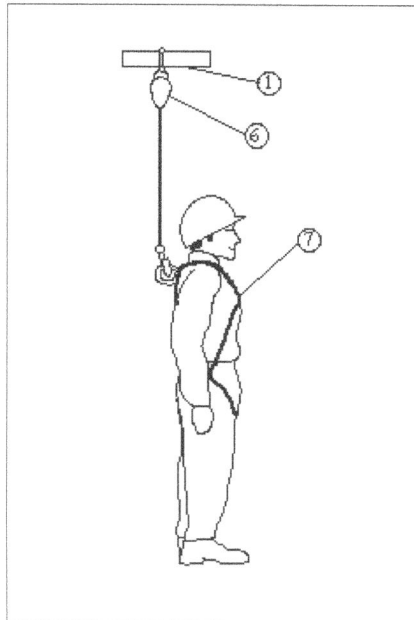

Figure 6-4. Components of Fall Protection Systems (2 of 4)

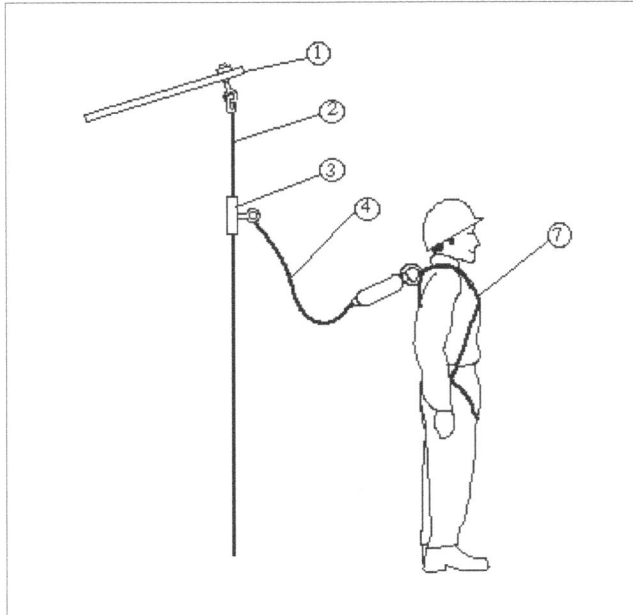

Figure 6-5. Components of Fall Protection Systems (3 of 4)

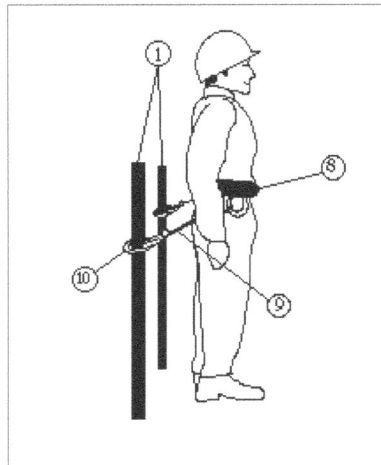

Figure 6-6. Components of Fall Protection Systems (4 of 4)

HEAD PROTECTION

HARD HATS

6-48. Hard hats are designed to provide protection to the head during exposures to potential hazards such as falling objects, striking against low hanging objects, or electrical hazards. Hard Hats will be worn during all cargo transfer operations, the unit commander or OIC in charge of the operation can designate the "KEVLAR" helmet as a substitute for the hard hat. **In general, protective helmets, or hard hats, should:**

- Resist penetration by objects
- Absorb the shock of a blow
- Be water resistant and slow burning
- Be provided with instructions explaining proper adjustment and replacement of the suspension and headband

6-49. Hard hats require a hard outer shell and a shock-absorbing lining. The lining should incorporate a head band and straps that suspend the shell from 1 to 1 1/4 inches (2.54 cm to 3.18 cm) away from the user's head. This design provides shock absorption during impact and ventilation during wear. The basic design of the hard hat is shown in Figure 6-7.

Figure 6-7. Hardhat Basic Design

Inspection

6-50. Head protection should be inspected prior to use and should be removed from service if the suspension system shows signs of deterioration such as:

- Cracking
- Tearing
- Fraying

The suspension system no longer holds the shell from 1 inch to 1 1/4 inches (2.54cm - 3.18cm) away from the wearer's head.

6-51. Inspect the outer surface and remove the hat from service if any of the following signs of deterioration are evident:

- The brim or shell is cracked, perforated, or deformed.
- The brim or shell shows signs of exposure to heat, chemicals, ultraviolet light, or other radiation. Such signs include:
  - Loss of surface gloss
  - Chalking
  - Flaking (This is a sign of advanced deterioration.)

Care Notes

6-52. Paints, paint thinners, and some cleaning agents can weaken the shell of the hard hat and may eliminate electrical resistance. Consult the helmet manufacturer for information on the effects of paint and cleaning materials on their hard hats. Keep in mind that paint and stickers can also hide signs of deterioration in the hard hat shell. Limit their use.

6-53. Ultraviolet light and extreme heat, such as that generated by sunlight, can reduce the strength of the hard hats. Therefore hard hats should not be stored or transported on the rear-window shelves of automobiles or otherwise in direct sunlight.

6-54. To clean, immerse for one minute in hot (approximately 140° F, or 60° C) water and scrub using detergent. Rinse in clear, hot water.

GANGWAYS

6-55. Whenever practicable, a gangway of not less than 20 inches (.51 m) in width, of adequate strength, maintained in safe repair and safely secured shall be used. Each side of the gangway, and the turntable, if used, shall have a handrail with a minimum height of 33 inches (.84 m) measured perpendicularly from rail to walking surfaces at the stanchion, with a mid-rail. Rails shall be of wood, pipe, chain, wire, rope or materials of equivalent strength and shall be kept taut always. Portable stanchions supporting railings shall be supported or secured to prevent accidental dislodgement. The gangway shall be kept properly trimmed. When a fixed flat tread accommodation ladder is used, and the angle is low enough to require soldiers to walk on the edge of the treads, cleated duckboards shall be laid over and secured to the ladder. When the gangway overhangs the water so that there is danger of soldiers falling between the ship and the dock, a net or suitable protection shall be provided to prevent soldiers from receiving serious injury from falls to a lower level. If the foot of a gangway is more than one foot (.30 m) away from the edge of the apron, the space between them shall be bridged by a firm walkway equipped with a hand rail with a minimum height of approximately 33 inches (.84 m) with mid-rails on both sides. Gangways shall be kept clear of supporting bridles and other obstructions, to provide unobstructed passage. If, because of design, the gangway bridle cannot be moved to provide unobstructed passage, then the hazard shall be properly marked to alert soldiers of the danger. Obstructions shall not be laid on or across the gangway. Handrails and walking surfaces of gangways shall be maintained in a safe condition to prevent soldiers from slipping or falling.

LOCK OUT / TAG OUT OF ENERGY SOURCES

6-56. Lockout/tagout procedures are necessary due to the complexity of modern Army watercraft and the cost and potential affects of delays associated with equipment down time. The procedure is also required due to the hazards to personnel which could result in their injury or in the worse case, death. This lockout/tagout program is mandatory for all Army watercraft. The program is designed to notify personnel that locked/tagged equipment or systems are NOT to be operated under any circumstances. The lockout/tagout system consists of a series of locks and tags that are attached to individual components to indicate that they are restricted from operation (red danger tag). Each tag contains the necessary information to prevent a possible injury to personnel or damage to installed equipment. Tags associated with the tagout procedure should never be used for valve identification, for marking leaks, or any other purpose not specified in this tagout procedure. Locks must be placed on the initial energy source (whether electrical, air, or fluid) nearest the item to be tagged. Use of locking devices is mandatory so that equipment needing or under repair/replacement can not be energized.

NOTE: All Class Bravo (B) and Charlie (C) vessels, without 20 level certificated personnel, will comply with vessel support office (VSO) SOP.

6-57. The use of tags or other labels is not a substitute for other safety measures such as chaining or locking valves, removing fuses, or racking out circuit breakers. However, tags will be attached to the fuse panel, racked out circuit breaker cabinet or locked valve, to indicate the need for such action. If any component has more than one type of tag or sticker attached; the DANGER (RED) tag, when present, will take precedence over all other tags or stickers. Standard lockout/tagout procedures are to be used for all maintenance, including work to be done by support maintenance units and local contractors. Each maintenance action will require its own set of tags even if two or more maintenance actions require the same equipment to be tagged. Never rely on the tags from other maintenance actions to provide protection for the work you are assigned to do. Lockout/tagout procedures will be enforced at all times. Violation of any tag compromises the entire tagout system and could in itself have serious consequences. Therefore, strict adherence to the tagout procedure without exception is required by all personnel. Lockout/tagout training should be conducted for all crewmembers to insure they are familiar with all procedures and meaning of various tags.

PROGRAM RESPONSIBILITIES

6-58. The following are the responsibilities of the Vessel Master, Chief Engineer, company commander, and company and/or battalion marine maintenance officer.

- Vessel Master: The Vessel Master is responsible for ensuring these procedures are properly utilized aboard his/her assigned watercraft.
- Chief Engineer: Aboard Army watercraft, the chief engineer is the Authorizing Officer. Insures all crewmembers are familiar with lockout/tagout procedures and tag meanings.
- Company Commander: The company commander is responsible for ensuring his/her unit is in compliance with these procedures. The commander ensures that these procedures are addressed within the company maintenance SOP to include the proper indoctrination of new unit personnel.
- Company and/or Battalion Marine Maintenance Officer: The company/battalion marine maintenance officer will routinely audit individual vessels to ensure compliance with this program.

APPLICATION

6-59. These requirements apply to all maintenance actions performed aboard Army watercraft where the unexpected energizing, startup, or release of stored energy of equipment would be likely to endanger personnel or the equipment itself.

6-60. The following describes terms related to lockout/tagout procedures.

- Authorizing Officer: The Authorizing Officer for lockout/tagout on Army watercraft will be the Chief Engineer.
- Energy Isolating Device: A mechanical device that physically prevents the release or transmission of energy. These devices include but are not limited to the following: Manually operated breakers, disconnects, or switches, valves, blank flanges.

NOTE: Push buttons, selector switches, and other types of circuit devices ARE NOT energy isolating devices.

- Energy source: Any device, component, or system which contains potential energy capable of injuring personnel or damaging installed equipment. Energy sources may be electrical, pneumatic, hydraulic, thermal, chemical, or in a mechanical form such as a rotating element.
- Lockout device: A device that uses a positive means such as a lock to hold an energy isolating device in a safe position preventing the energizing of equipment or the release of another form of energy. Lockout devices include any device which mechanically prevents the energy isolating device from being repositioned. This may be as simple as wire rope with clips.
- Maintenance action: Any preventative or corrective maintenance performed by the vessel's crew and support unit maintenance personnel and private contractor personnel. Each maintenance action will require its own set of tags. This does not include maintenance performed during a cyclic maintenance period. Lockout/tagout will be governed by the shipyard performing the cyclic maintenance.

- Tagout: Tags affixed to energy isolating devices for warning purposes. They DO NOT provide the physical restraint that lockout devices provide. These tags are as follows:
  - Danger tag: This tag is red. It prohibits the operation of equipment that could jeopardize the safety of personnel or endanger equipment and associated systems. Under no circumstances will equipment be operated when danger tags are attached. Laminated danger tags intended for repeated use will not be used aboard Army watercraft.
  - Caution tag: This tag is yellow It is used as a precaution to advise personnel of temporary special instructions or to indicate that unusual caution must be exercised to operate equipment. These instructions must state the specific reason why the tag is installed. The phrase "Do not operate without the Chief Engineer's permission" is not acceptable since no equipment should be operated without direct permission.
- Tagout log: The tagout log consists of the Tagout Index, active Tagout Record Sheets, and the inactive Tagout Record Sheets. The Tagout Index and Tagout Record provides a means of tracking actions, ensures that serial numbers are sequentially issued, and assists in conducting audits and reviews of the tagout program for the vessel. It also provides a ready means of referral for the crew. The sheet may be locally reproduced.

PROCEDURES

6-61. The following describes lockout/tagout procedures to be followed by all Army watercraft personnel.

6-62. Preparation of Tags and Logs: Danger and Caution tags

- Each tagout action is assigned a serial number in sequence from the Tagout Log Index. This serial number will also be used to identify each tag associated with the tagout action. When a tagout action requires more than one tag, the same base number will be used with a sequence number to identify individual tags (for example, 001-1, 001-2, and so on).
- Tagout entries will provide sufficient information to give personnel reviewing the log a clear understanding of the purpose and necessity for each tagout action.
- Enough tags and lockout devices will be used to completely isolate the system or component being worked on and to prevent operation of a system or component from all stations that could exercise control. System diagrams and circuit schematics should be used to determine the adequacy of all tagout actions.
- The person requesting the tags will prepare the Tagout Record Sheet and associated tags.
- The Authorizing Officer (normally the Chief Engineer) will review the tagout log entries and tags for completeness and accuracy. When satisfied, the Authorizing Officer will sign the Tagout Record Sheet and tags authorizing the installation of the tags.
- The person attaching the tags and lockout devices will ensure that the items being tagged are in the prescribed position or condition (for example, shut, locked shut, fuses removed, and so on) exactly as stated on the tag, then sign and attach the tags and locking devices if required. The tags will be securely attached so they are apparent to anyone who may try and operate the component.
- After the tag is attached, a second person will independently verify the tagged equipment or component is in the position or condition indicated on the tag and that the tag and lockout device, if required, is properly attached. That person will sign the tags and the Tagout Record Sheet. Only qualified personnel will perform the second check of tag installation.
- Locking devices and tags are only valid for 30day increments. One of three occurrence must happen at the end of this time frame:
  - Equipment is repaired and returned to normal usage. Devices and tags are removed and tagout record sheet properly annotated.
  - The identified system, locking devices, and tags are inspected by the same individuals who initiated the process to ensure validity. Tags and tagout record sheet are annotated with new date.
  - When in initiating individuals are not available, the original devices and tags are removed and the tagout record sheet reflects the entry being "cleared". This procedure causes the lock out/tag

out process to begin from the beginning with a new equipment serviceability inspection and the issuing of new lock out/tag out with new date and assignment number.

6-63. Tag Removal. Danger and Caution tags will be removed immediately when the situation requiring the tagout has been corrected. Danger tags will be properly cleared and removed before a system or portion of a system is operationally tested and restored to service.

- No tags and lockout devices will be cleared without the approval of the Authorizing Officer. The Authorizing Officer's approval will be annotated on the Tagout Record Sheet indicating which tags are to be removed.
- The person who initiated the tagout should, if possible, be the person who clears the tags. However, do not delay the removal of tags when work is completed and the individual is not available.
- As the tags and lockout devices are removed, they will be returned immediately to the Authorizing Officer. Using the returned tags and Tagout Record Sheet, the Authorizing Officer will verify that all the tags have been cleared by all parties. The date and time cleared will be annotated on the Tagout Record Sheet and the date entered on the Tagout Index Sheet.
- Removed tags will be destroyed after they have been delivered to the Authorizing Officer. The Authorizing Officer will file it in the inactive section of the Tagout Log. Inactive Tagout Record Sheets will be maintained for six months, and then destroyed.

6-64. Lost or Missing Tags. Tags which are missing or have come off the item to which they are to be attached will immediately be reported to the Authorizing Officer.

- The Authorizing Officer will direct a new tag to be added to the Tagout Record Sheet using the above procedure for initiating a tagout.
- After the new tag is installed and verified by a second party, the Authorizing Officer will then clear the old tag from the Tagout Record Sheet using the above procedures for clearing/removing tags.

This table describes the DANGER tag

TAG	COLOR	FUNCTION	Operate The Component?
DANGER	RED	Prevent equipment from being energized	NO

PROCEDURES FOR COMPLETING THE DANGER TAG

Step	Action	By Whom	Location On Tag
1	Select the proper tag. (Example: Danger, Caution)	Person doing work	
2	Fill in the system comp. identification. (Example: #1 SSDG)	Person doing work	A
3	Fill in date/time.	Person doing work	B
4	Indicate position or condition of item being tagged.	Person doing work	C
5	Enter serial # in num. order from Tagout Index.	Person doing work	E
6	Authorizing Officer signs tag.	Chief Engineer	F
7	Position the valve or component, install locking devices if required. Sign tag and hang.	Person doing work	D
8	Second person performs independent check and signs tag.	Person checking tags	B

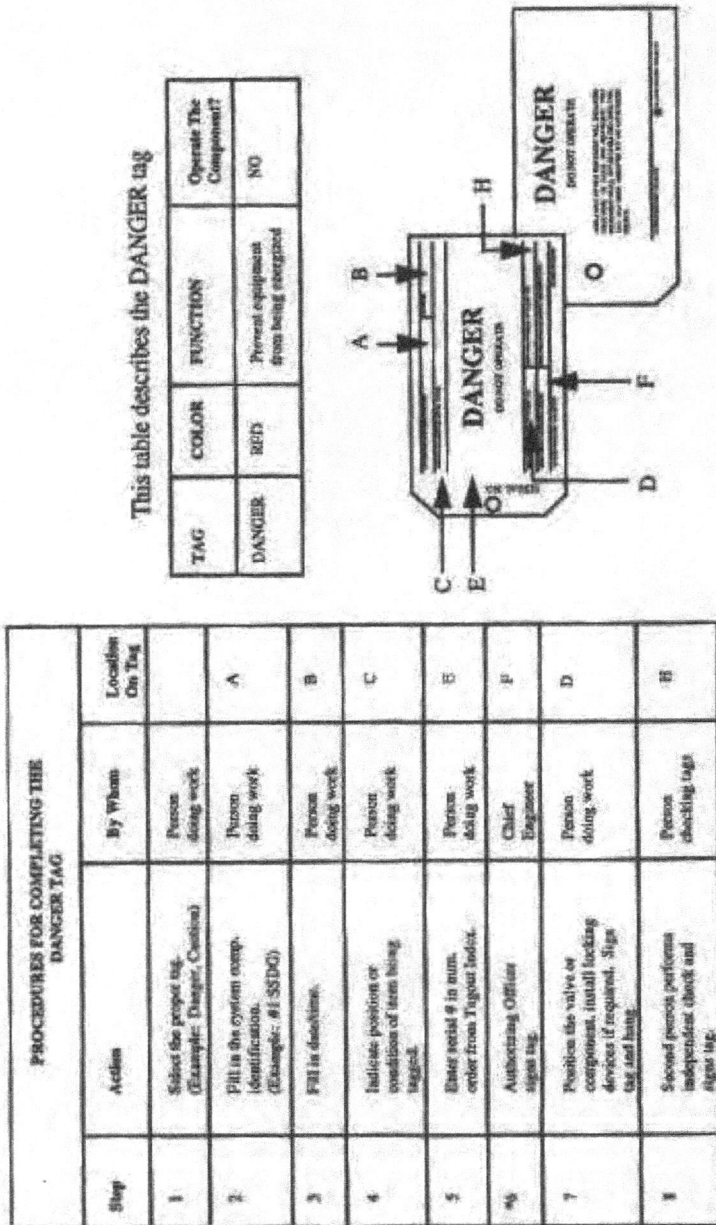

*Before the Authorizing Officer signs the tag, the Tagout Record Sheet must be completed by the person isolating the tagout.

Figure 6-8. Danger tag

This table describes the DANGER tag

TAG	COLOR	FUNCTION	Operate The Component?
DANGER	RED	Prevent equipment from being energized	NO

DANGER
DO NOT OPERATE

DANGER
DO NOT OPERATE

PROCEDURES FOR COMPLETING THE DANGER TAG			
Step	Action	By Whom	Location On Tag
1	Select the proper tag. (Example: Danger, Caution)	Person doing work	
2	Fill in the system comp. identification. (Example: #1 SSDG)	Person doing work	A
3	Fill in decelines	Person doing work	B
4	Indicate position or condition of item being tagged	Person doing work	C
5	Enter serial # in num. order from Tagout Index.	Person doing work	E
6	Authorizing Officer sign tag.	Chief Engineer	F
7	Position the valve or component, install locking devices if required. Sign tag and hang.	Person doing work	D
8	Second person perform independent check and sign tag.	Person checking tags	H

*Before that Authorizing Officer signs the tag, the Tagout Record Sheet must be completed by the person initiating the tagout.

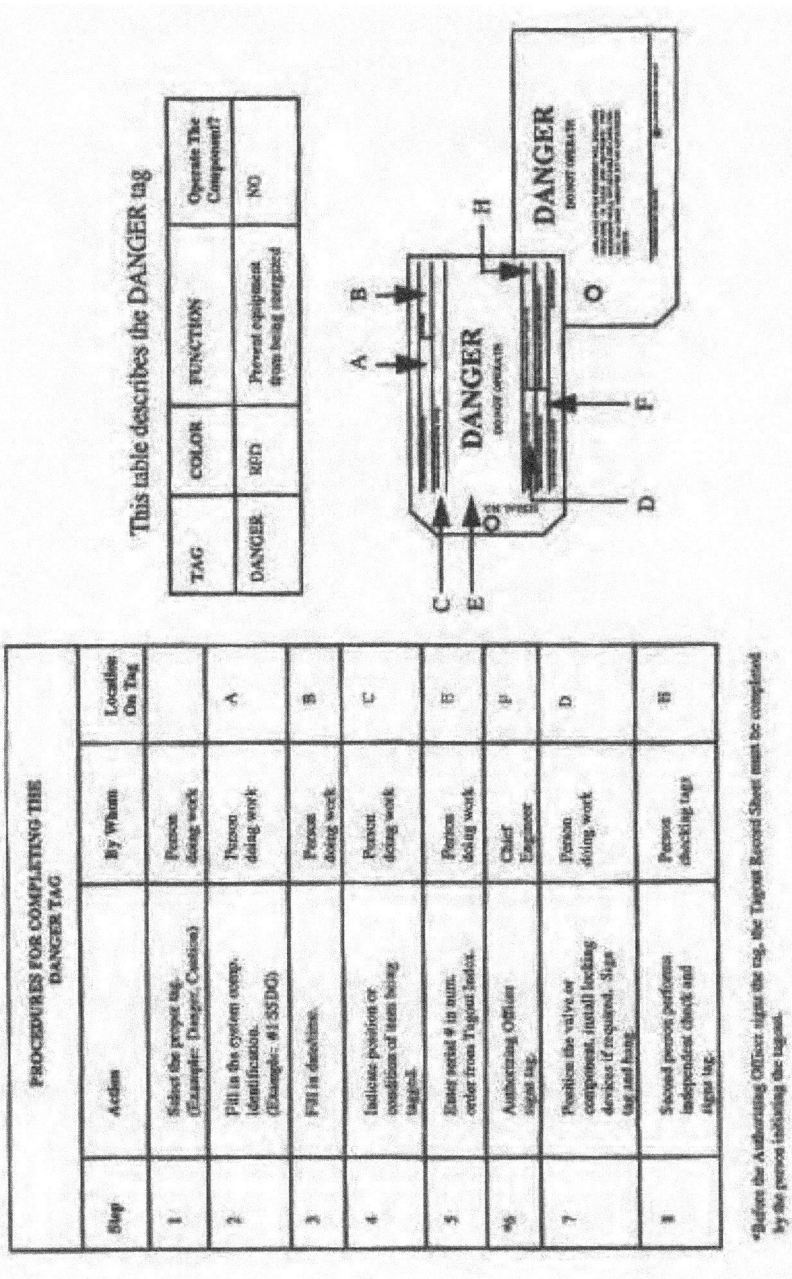

Figure 6-9. Caution tag

PROCEDURES FOR TAGOUT RECORD SHEET LOCATION	
Log serial number. Each tagout is assigned a log serial number in sequence. The Tagout Index will be used for assigning log serial numbers. Enter the system or component being tagged out.	A
The reason for the tagout, the hazards involved, amplifying instructions, and work necessary to clear the tags will be sufficiently detailed to give watch standers reviewing the Tagout Log a clear understanding of the purpose of and necessity for each tagout action.	B
Enough tags should be used to completely isolate the system, piping, or circuit being worked on or to prevent operation of a system. Each tag will contain the Log serial number followed by a dash and number (Example: 001-1). The -1 indicated the first tag for serial number 001. Additional tags would be numbered -2, -3, and so on.	C
The location (for example #1 SSDG circuit breaker) and the position/condition (for example: open, shut, locked shut, racked out, fuses removed, and so on) of the tagged item should be indicated by the most easily identifiable means. The position/condition column need not be filled in for caution tags.	D
The Authorizing Officer will review the Tagout Record Sheet and tags for completeness and accuracy. When satisfied, he/she will sign and enter the date/time the Tagout Record Sheet and tags authorizing the tags to be installed.	E
The individual installing the tags and locking devices will reposition the item to conform to the required position / condition as stated on the tag. He/she will then sign the tag and then hang it. This same individual will then initial the Tagout Record Sheet indicating the tag and locking devices were installed.	F
After the person initiating the tagout has completed installing the tags, a second individual will independently check each tag to ensure the tag and any locking devices are properly installed and the item is in the proper condition/position. This individual will sign the tag and then initial the Tagout Record Sheet to indicate the tag was correctly installed.	G
After the work is completed, the Authorizing Officer (Chief Engineer) will inspect and when satisfied, authorize removal of the tags by signing the Tagout Record Sheet.(exception is class B and C vessels without 20 level certified engineer on board)	H
The individual assigned to clear the tags and locking devices will enter the date/time the tag was cleared and initial the Tagout Record Sheet indicating the tag was removed. All removed tags will be returned to the Authorizing Officer for destruction. The Tagout Record Sheet will be filed in the inactive section of the Tagout Log for six months.	I

SIGNAGE

6-65. On Army vessels, two general types of signs are used: safety and photo-luminescent.

6-66. Safety signs are posted for protect personnel from hazardous situations or areas. Some signs are posted at entry ways while others are posted on specific equipment to warn of hazardous activities or to inform them of specific equipment operation steps. The list below is not all-inclusive.

6-67. Some required entry signs are:
- Hearing Protection Required
- Escape Hatch, Do Not Block
- Keep Closed
- No Smoking
- No Flame
- High Radiation Hazard
- Confined Space, Do Not Enter
- Exit

- Arrows (Directional)

6-68. Required specific equipment signs:

- Vision Protection Required
- Wear Protective Clothing (gloves, apron, etc)
- No Smoking
- High Radiation Hazard
- Fixed Fire Fighting System Diagrams
- Emergency Steering Diagrams

6-69. Photo-luminescent signs are used to identify emergency escape routes, hatches, and some emergency equipment locations. Do not use these sign on exterior locations. The listing below is not inclusive of all requirements.

- Emergency Escape Breathing Device (EEBD)
- Exit
- Exit (with direction arrow)
- "Escape Hatch (or Scuttle), Do Not Block"

6-70. Direction Arrow - Direction arrows can be made using the tape and cutting pointed ends. These direction arrows are required to be placed between seven (7) and twelve (12) inches from the floor in all interior passageways. Additionally, these arrows can be placed in other spaces to expedite emergency egress from those areas.

FIRST AID KIT, BURN TREATMENT

NOTE: This section contains information on the new federally authorized first aid burn dressing. Throughout this section the burn dressing will be called or referred to by its commercial name "Water-Jel".

6-71. Water-Jel is a unique multi-use product for emergency burn care and fire protection. This patented product is designed to help save lives, increase the relief of pain and suffering, and reduce tissue damage caused by burns. Water-Jel products have been adopted into the Federal Supply System by the Defense Medical Standardization Board and approved for sale by the US Food and Drug Administration.

6-72. Water-Jel is a one-step system that combines a scientifically formulated gel and a special carrier material. The gel is biodegradable, bacteriostatic, and water soluble. Water-Jel can be carried anywhere. When placed on the burn victim, it extinguishes the flames and immediately lowers and stabilizes skin temperature, helping to ease the pain and calm the victim. Because the product is bacteriostatic, the covered wound is protected from further contamination. In addition to providing essential burn care, Water-Jel performs other lifesaving tasks. The product is water soluble, making removal of burnt clothing and jewelry easier. Water-Jel may also be used by a rescuer to shield himself and the victim from the intense heat and flames of a fire. In its larger sizes, Water-Jel may be used to extinguish a small fire and help provide a means of escape from larger fires. Water-Jel comes in a variety of sizes, from a 6-foot X 5-foot fire blanket to a 4-inch X 4-inch sterile burn dressing. The convenient packet or containers can be easily carried in all types of vessels and vehicles. They can also be stored in areas that are readily accessible to rescuers or burn victims.

6-73. The Burn Treatment First Aid Kit includes the following:

QUANTITY (UI)	DIMENSIONS	TYPE
2 each	8" x 18"	Dressing
1 each	4" x 16"	Dressing
4 each	4" x 4"	Dressing
1 each	12" x 16"	Dressing
1each	Burn-Jel Kit	Topical Dressing
1each	72" x 60"	Blankets
1 each	VCR Tape	Informational

TECHNICAL SPECIFICATIONS

6-74. The following describes the technical aspects of Water-Jel.

- Appearance- Off White Translucent
- Odor- Characteristic Menthol
- Environmental concerns- Biodegradable

EXPIRATION DATES

6-75. Burn dressings sealed in a foil package expire three years from the date of manufacture. The manufacture date is part of the lot number stamped into the edge of the package. Burn blankets and wraps in containers expire five years from the date of manufacture. The manufacture date is part of the lot number located on a white sticker fastened to the container. Lot numbers read: DAY/MONTH/YEAR.

PLACEMENT

6-76. First aid kits should be strategically placed upon the vessels that would render them most effective in case of an emergency. Before placing kits into service, they must be inspected and inventoried. The inventory checklist will be posted on the outside of the kit annotating all items that have a specific shelf life, along with the expiration date of that item. The kits should also be sealed with some type of tamper seal, but not to hinder the entry by persons in need. The kit must have an inventory conducted (at a minimum) annually.

Itemized Safety Equipment

Legend
N/A – Not applicable
R – As required
Italics – Additional
pieces, components,
storage or mounting

Table A-1. Itemized Safety Equipment

Various sizes of worn items may be used to meet total requirement

Nomenclature and NSN	TSV	LSV	LT 128	LT100	LCU 2000	LCU 1600	ST	LCM MOD1	LCM MOD2	BD 89	BD 115
Life Saving Equipment											
Anti-Exposure Coveralls Small: 8410-01-011-5051 Medium: 8410-01-011-5052 Large: 8410-01-011-5053 Extra Large: 8410-01-011-5054	10	10	11	10	6	6	12	4	4	15	18
Emergency Escape Breathing Device 4240-01-439-5937	44	44	27	16	22	22	5	1	1	15	20
Flotation Kit with Hoisting Sling 6530-01-370-6905 or 6530-01-459-7983 *Flotation Kit only*: 4220-01-329-6420 *Hoisting Sling only*: 1670-01-226-5300 *Bar, Ballast Kit*: 4220-01-517-0088 *Flotation Logs*: 6530-01-421-9334 *Log Covers*: 6530-01-459-4158 *Chest Pads*: 4240-01-533-5040 *Chest Pad Covers*: 4240-01-533-5038	2	2	2	1	1	1	1	NA	NA	1	1
Heaving Line, Safety w/ ball 4020-01-344-0552 / 4020-01-387-8795	8	8	4	4	4	4	4	2	2	4	4
Immersion Suits	41	36	30	18	23	17	12	3	3	NA	NA

Table A-1. Itemized Safety Equipment

Various sizes of worn items may be used to meet total requirement

Nomenclature and NSN	TSV	LSV	LT 128	LT100	LCU 2000	LCU 1600	ST	LCM MOD1	LCM MOD2	BD 89	BD 115
Adult, Universal 4220-01-251-6466 Adult, Jumbo 4220-01-251-6467											
Life Preserver, Type I, Deluxe 4220-01-485-1135	312	40	30	18	23	20	12	6	6	15	15
Life Raft, Commercial	4	NA	NA	NA	NA	NA	1	NA	NA	NA	NA
Flashlight, Watertight 6230-00-264-8261	4	4	4	4	4	2	2	1	1	4	4
Life Raft, Navy Mk7 1940-01-015-7346	NA	4	2	2	2	1	NA	NA	1	1	1
Light Marker, Distress: 6230-01-143-4778 Battery, 6v: 6135-00- 100-0413	6	6	6	5	6	5	3	2	2	4	4
Line Throwing Device, 45-70 1095-00-270-6019	1	1	1	1	1	1	1	NA	NA	NA	NA
Cartridges, 45-70 1095-00-240-7164 (BX)	1	1	1	1	1	1	1	NA	NA	NA	NA
Litter, Semi-rigid, Pole-less, Neil Robertson: 6530-00-783-7600	1	1	1	1	1	1	1	NA	NA	1	1
Marker, Location Marker, Marine Mk58 1370-00-028-6010, PN 78-0-68	2	2	2	2	2	1	NA	NA	NA	NA	NA
MEDEVAC II Litter 6530-01-338-6094 Straps, Patient Restraint (5 ea): 6530-01-419-3379 Strap, Retainer, 4ea: 5340-01-421-7635 Liner, Plastic Net (LSC P/N 439): 6530-02-407-1970 Cover, Litter: 6530-01-407-6015	2	2	2	1	1	1	1	NA	NA	1	1
Personnel Marker Lights,	R	R	R	R	R	R	R	R	R	R	R

Table A-1. Itemized Safety Equipment

Various sizes of worn items may be used to meet total requirement

Nomenclature and NSN	TSV	LSV	LT 128	LT100	LCU 2000	LCU 1600	ST	LCM MOD1	LCM MOD2	BD 89	BD 115
Chemical 6260-01-086-8077											
Reflective Tape 3" X 50 yds: (Roll) 9390-01-078-8660	1	1	1	1	1	1	1	1	1	1	1
Release Lifesaving Unit (hydrostatic) 4220-01-493-9233	R	R	R	R	R	R	R	NA	R	R	R
Ring Buoy, 30" 4220-00-275-3157 *Retrieving Line, Orange 1", 200 yds* 4020-00-530-0698	8	8	8	8	8	6	4	3	3	5	5
Signal Illumination, Ground, red star para 1370-00-629-2336, P/N 8797968	12	12	12	12	12	12	NA	NA	NA	NA	NA
Signal, Smoke and Illumination, Mk124 1370-00-092-9921	NA	NA	NA	NA	NA	NA	12	12	12	12	12
Signal, Whistle, Plastic Ball 8465-00-254-8803 *Cord, Nylon, Type I*: 4020-00-240-2145	R	R	R	R	R	R	R	R	R	R	R
Sling, Litter Hoisting 1670-01-226-5300	2	2	1	1	1	1	1	NA	NA	1	1
Snap, halyard: 5340-00-275-4584	R	R	R	R	R	R	R	R	R	R	R
Talc, non-allergic powder 6810-00-270-9989	R	R	R	R	R	R	R	R	R	R	R
Training EEBD 4240-01-459-0078	1	1	1	1	1	1	1	NA	NA	1	1
Work Vests, Type III 4220-01-415-9817	15	15	10	10	6	6	5	6	6	6	6
Fire Fighting											
Fire Axe, 4210-00-142-4949	14	14	6	4	6	4	1	1	1	1	1
Foam, liquid, AFFF 5 Gal	12	12	6	6	8	8	3	NA	NA	NA	NA
Hose, Fire, 1 ½"x 50': 4210-01-131-0249	28	28	7	4	10	6	4	NA	NA	5	5
Hose, Fire, 2 ½"x50':	6	6	4	4	4	NA	NA	NA	NA	NA	NA

Table A-1. Itemized Safety Equipment

Various sizes of worn items may be used to meet total requirement

Nomenclature and NSN	TSV	LSV	LT 128	LT100	LCU 2000	LCU 1600	ST	LCM MOD1	LCM MOD2	BD 89	BD 115
4210-01-131-0247											
Nozzle, Applicator, 10 foot 4210-00-372-0865	3	3	2	1	2	2	1	NA	NA	2	2
Nozzle, Applicator, 4 foot 4210-00-372-0864	5	5	2	1	3	2	1	NA	NA	2	2
Nozzle, Fire Foam, Aeration, 1 1/2 inch, Local purchase, Mod 3951, NPSH thread	6	6	2	1	NA	2	NA	NA	NA	NA	NA
Nozzle, Fire, 1 1/2 inch, 3 position 4210-00-392-2943	15	15	4	2	4	5	4	NA	NA	2	2
Nozzle, Fire, 1 1/2 inch, Variable 4210-00-465-1906	2	2	4	2	4	NA	NA	NA	NA	2	2
Nozzle, fire, 2 1/2 inch, 3 position 4210-00-392-2944	5	5	NA	NA	NA	NA	NA	NA	NA	NA	NA
Portable Fire Extinguisher, 15 lb, CO2 4210-00-203-0217	R	R	R	R	R	R	R	R	R	R	R
Portable Fire Extinguisher, Dry Chem, 5 lb, 4210-00-775-0127	R	R	R	R	R	R	R	R	R	R	R
Portable Fire Extinguisher, Dry Chem, 10 lb, 4210-00-889-2491	R	23	R	R	19	R	7	R	R	R	R
Wrench, Spanner 5120-01-148-2422	R	17	R	R	R	R	R	R	R	R	R
Fire Ensemble											
Anti-Flash Gloves 8415-01-267-9661	12	12	29	16	6	6	5	6	6	15	15
Anti-Flash Hood 8415-01-268-3473	12	12	29	16	6	6	5	6	6	15	15
Firefighter Coveralls, required sizes (any combination) Sm Reg 4210-01-468-5528 Med Reg 4210-01-468-5551 Lg Reg 4210-01-468-5565	12	12	7	5	6	6	4	NA	NA	4	4

Table A-1. Itemized Safety Equipment

Various sizes of worn items may be used to meet total requirement

Nomenclature and NSN	TSV	LSV	LT 128	LT100	LCU 2000	LCU 1600	ST	LCM MOD1	LCM MOD2	BD 89	BD 115
XLg Reg 4210-01-468-5671											
Self Contained Breathing Apparatus 4240-01-545-9605	16	16	9	5	9	6	4	NA	NA	4	4
Breathing Cylinders See SCBA TM	32	32	18	10	18	12	8	NA	NA	8	8
Firefighter's Gloves (any combination) Med 4210-01-335-7902 Lg 4210-01-335-7903 XLg 4210-01-335-7904	12	12	7	5	6	6	4	NA	NA	4	4
Firefighter's Helmet 4210-01-493-7428	12	12	7	5	6	6	4	NA	NA	4	4
Firemen's Boots, various sizes Size 8: 8430-00-753-5938 Size 10: 8430-00-753-5940 Size 12: 8430-00-753-5942	12	12	7	5	6	6	4	NA	NA	4	4
Flashlight, Explosion Proof, 2 cell 6230-00-269-3034	12	12	7	5	6	6	4	NA	NA	4	4
Kit, bag, flyers 8460-00-606-8366	12	12	7	5	6	6	4	NA	NA	4	4
Wire Rope Life Line (Tending) 4010-00-285-9901	12	12	7	5	2	2	4	NA	NA	6	6
Damage Control											
Adapter, Vent Hose 4730-01-378-5288	2	2	1	1	1	1	NA	NA	NA	1	1
Eductor, bilge, 4 inch 4320-00-256-8206	2	2	1	1	1	1	1	NA	NA	1	1
Electrical Tool Kit 5180-00-391-1087	1	1	1	1	1	NA	1	NA	NA	1	1
Exhaust Hose, Engine 4210-00-776-0657	2	2	1	1	1	1	1	NA	NA	1	1
Fan, Water-Driven 4140-01-333-2224	2	2	1	1	1	1	NA	NA	NA	1	1
Foot Valve w/Strainer 4820-00-540-2381	2	2	1	1	1	1	1	NA	NA	1	1

Table A-1. Itemized Safety Equipment

Various sizes of worn items may be used to meet total requirement

Nomenclature and NSN	TSV	LSV	LT 128	LT100	LCU 2000	LCU 1600	ST	LCM MOD1	LCM MOD2	BD 89	BD 115
Hose, Assembly, Air Duct 4720-00-277-7225	2	2	1	1	1	1	NA	NA	NA	1	1
Hose, Discharge, 4 inch 4210-01-220-6648	2	2	1	1	1	1	1	NA	NA	1	1
Hose, Suction 4210-00-725-9234	2	2	2	2	2	2	2	NA	NA	2	2
Lantern, Electric Battle, fixed	R	R	R	R	R	R	R	R	R	R	R
P-100, Dewatering Pump 4320-01-387-2869	2	2	1	1	1	1	1	NA	NA	1	1
Plug 1"x0"x3": 5510-00-260-8953	10	10	20	10	10	5	10	5	5	5	5
Plug 2"x0"x4": 5510-00-260-8958	10	10	20	10	10	5	10	NA	NA	5	5
Plug 3"x0"x8": 5510-00-260-8962	10	10	20	10	10	5	10	10	10	5	5
Plug 5"x1"x10": 5510-00-260-8966	10	10	20	10	NA	NA	NA	NA	NA	NA	NA
Pump, Centrifugal, Electrical, Submersible 4320-00-368-3186	2	2	2	NA	1	1	NA	NA	NA	NA	NA
Repair Kit, Pipe, Emergency Damage 4730-01-414-6976	1	1	1	1	1	1	1	NA	NA	1	1
Ship's Maul 5120-00-255-1476	2	2	1	1	1	1	1	1	1	1	1
Shoring batten, steel adjustable 4.5' to 7': 5210-01-418-4746	10	10	6	5	5	5	3	1	1	5	5
Shoring batten, steel adjustable 7' to 12': 5210-01-418-4748	10	10	6	5	5	5	3	1	1	5	5
Valve, Ball, Tri-Gate 4210-01-038-6001	2	2	1	1	1	1	1	NA	NA	1	1
Wedge, 1 1/2"x 2"x12" 5510-00-268-3475	5	5	10	5	10	5	10	6	6	5	5
Wedge, 1 1/2"x3"x12": 5510-00-268-3476	5	5	5	5	5	5	NA	6	6	5	5
Wedge, 2 ¼"x3"x18": 5510-00-268-3477	5	5	5	5	5	NA	NA	NA	NA	NA	NA
Wedge, 2"x2"x8": 5510-00-	5	5	10	5	10	5	NA	6	6	5	5

Table A-1. Itemized Safety Equipment

Various sizes of worn items may be used to meet total requirement

Nomenclature and NSN	TSV	LSV	LT 128	LT100	LCU 2000	LCU 1600	ST	LCM MOD1	LCM MOD2	BD 89	BD 115
268-3479											
Wedge, 2"x4"x8": 5510-00-268-3480	5	5	10	5	10	5	10	NA	NA	5	5
Wedge, 3"x3"x12": 5510-00-268-3481	5	5	10	5	10	5	10	NA	NA	5	5
Confined Space Entry											
Lantern, Battle, Portable 6230-01-141-2901	R	R	6	6	R	R	R	NA	NA	17	17
Meter, Detection, Gas: 6665-01-529-8483	1	1	1	NA	1	NA	NA	NA	NA	NA	NA
Sampling Pump, Hydrogen Fluoride Gas 6665-01-429-8592	1	1	1	NA	1	NA	NA	NA	NA	NA	NA
Tripod, Rescue System (Salalift II, model number 8300030)	1	1	NA	NA	1	NA	NA	NA	NA	NA	NA
General											
Battery Maintenance											
Apron, Utility: 8415-00-082-6108	1	1	1	1	1	1	1	1	1	1	2
Eyewash Location Sign: 9905-01-345-4521	2	1	2	1	1	1	1	1	1	1	1
Face-shield, Industrial, Clear Plastic Visor 4240-01-063-5996	4	4	1	5	NA	NA	NA	NA	NA	4	4
Fountain, Eye and Face Wash, Portable 4240-01-258-1245 (BX)	2	1	2	1	1	1	1	1	1	1	1
Fungicide Additive: 6840-01-267-4346	2	1	2	1	1	1	1	1	1	1	1
Gloves, Chemical: 8415-00-266-8677	1	1	1	1	1	1	1	1	1	1	1
Goggles, Industrial, Non-Vented 4240-00-190-6432	4	4	2	2	15	14	2	3	3	6	6
Personal Protective Equipment											
Blanket, WaterJel, 6'x5' 6510-01-242-2271	4	4	2	2	2	1	1	NA	NA	NA	NA

Table A-1. Itemized Safety Equipment

Various sizes of worn items may be used to meet total requirement

Nomenclature and NSN	TSV	LSV	LT 128	LT100	LCU 2000	LCU 1600	ST	LCM MOD1	LCM MOD2	BD 89	BD 115
Burn Treatment Water Jel, various	R	R	R	R	R	R	R	R	R	R	R
Face-shield, Industrial, Tilt: 4240-00-542-2048	3	3	NA	1	3	14	1	1	1	3	3
First Aid Kit, Burn Treatment 6545-01-526-9237	2	2	2	2	2	1	1	1	1	1	1
First Aid Kit, General, Ship 6545-00-116-1410	2	2	2	1	NA	2	NA	NA	NA	1	1
First Aid Kit, Life raft: 6545-00-168-6893	4	4	2	2	2	1	1	NA	NA	NA	NA
First Aid Kit, Small 12 unit 6545-00-922-1200	2	2	NA	2	2	NA	2	1	1	NA	NA
Gloves, Unisex: 8415-00-268-7871	2	2	2	2	2	2	2	2	2	2	2
Goggles, Industrial, Vented: 4240-00-052-3776	15	15	8	5	15	14	2	3	3	6	6
Hard Hats Orange: 8415-00-935-3136 White: 8415-00-935-3139	12	12	15	10	10	10	4	3	3	8	8
Harness, Safety, Torso: 4240-01-488-6995	4	4	4	2	6	4	2	2	2	2	2
Lanyard w/ Dyna Brake 4240-00-022-2521	4	4	4	2	4	NA	NA	NA	NA	2	2
Lanyard, Working: 4240-00-022-2518	4	4	4	2	6	4	2	2	2	2	2

Table A-2. Modular Causeway System Equipment

Nomenclature and NSN	FC	MWT MOD1	MWT MOD2	RRDF	SLWT	CF
Life Saving Equipment						
Anti-Exposure Coveralls Small: 8410-01-011-5051 Medium: 8410-01-011-5052 Large: 8410-01-011-5053 Extra Large: 8410-01-011-5054	18	8	8	18	8	8
Heaving Line, Safety w/ ball 4020-01-344-0552/ 4020-01-387-8795	NA	2	2	NA	2	2
Immersion Suits Adult, Universal 4220-01-251-6466 Adult, Jumbo 4220-01-251-6467	18	8	8	18	8	8
Life Preserver, Type I, Deluxe 4220-01-485-1135	NA	8	8	NA	8	8
Flashlight, Watertight 6230-00-264-8261	6	2	2	2	2	2
Light Marker, Distress: 6230-01-143-4778 *Battery, 6v*: 6135-00- 100-0413	4	2	2	4	2	2
Personnel Marker Lights, Chemical 6260-01-086-8077	R	R	R	R	R	R
Reflective Tape 3" X 50 yds: (Roll) 9390-01-078-8660	NA	1	1	NA	1	1
Retrieving Line, Orange 1", 200 yds 4020-00-530-0698	1	1	1	1	1	1
Ring Buoy, 30" 4220-00-275-3157	4	2	2	4	2	2
Signal, Smoke and Illumination, Mk124 1370-00-092-9921/ 1570-01-030-8330	NA	12	12	NA	12	12
Signal, Whistle, Plastic Ball 8465-00-254-8803 *Cord, Nylon, Type I*: 4020-00-240-2145	R	R	R	R	R	R
Snap, halyard: 5340-00-275-4584	R	R	R	R	R	R
Talc, non-allergic powder 6810-00-270-9989	R	R	R	NA	R	R
Work Vests, Type III 4220-01-415-9817	18	8	8	18	8	8

Table A-2. Modular Causeway System Equipment

Nomenclature and NSN	FC	MWT MOD1	MWT MOD2	RRDF	SLWT	CF
Fire Fighting						
Fire Axe	2	1	1	2	1	1
Portable Fire Extinguisher, 15 lb, CO2 4210-00-203-0217	NA	3	3	NA	NA	3
Portable Fire Extinguisher, Dry Chem, 5 lb, 4210-00-775-0127	NA	NA	NA	NA	6	NA
Portable Fire Extinguisher, Dry Chem, 10 lb, 4210-00-889-2491	2	NA	NA	2	2	NA
Fire Ensemble						
Anti-Flash Gloves 8415-01-267-9661	18	6	6	18	6	6
Anti-Flash Hood 8415-01-268-3473	18	6	6	18	6	6
Damage Control						
Plug 1"x0"x3": 5510-00-260-8953	NA	5	5	NA	5	5
Plug 2"x0"x4": 5510-00-260-8958	NA	5	5	NA	NA	5
Plug 3"x0"x8": 5510-00-260-8962	NA	5	5	NA	5	5
Plug 7"x3"x10": 5510-00-260-8969	NA	5	5	NA	5	5
Plug 8"x4"x10": 5510-00-260-8973	NA	5	5	NA	NA	5
Plug 10"x7"x12": 5510-00-260-8949	NA	5	5	NA	NA	5
Repair Kit, Pipe, Emergency Damage 4730-01-414-6976	NA	1	1	NA	1	1
Ship's Maul 5120-00-255-1476	NA	1	1	NA	1	1
Shoring batten, steel adjustable 3' to 5': 2090-00-058-3737	NA	4	4	NA	4	4
Wedge, 1 1/2"x 2"x12" 5510-00-268-3475	NA	5	5	NA	5	5
Wedge, 2"x2"x8": 5510-00-268-3479	NA	5	5	NA	5	5
Lantern, Battle, Portable 6230-01-141-2901	NA	3	3	NA	3	3
Confined Space Entry						
Meter, Gas Volume	1	1	2	1	NA	1
General						
Battery Maintenance						
Apron, Utility: 8415-00-082-6108	NA	2	2	2	2	2

Table A-2. Modular Causeway System Equipment

	Nomenclature and NSN	FC	MWT MOD1	MWT MOD2	RRDF	SLWT	CF
	Eyewash Location Sign: 9905-01-345-4521	NA	1	1	1	1	1
	Fountain, Eye and Face Wash, double station, ind. squeeze bottle	NA	1	1	1	1	1
	Gloves, Chemical: 8415-00-266-8677	6	2	2	6	2	2
	Goggles, Industrial, Non-Vented 4240-00-190-6432	NA	2	2	2	2	2
Personal Protective Equipment							
	Blanket, WaterJel, 6'x5' 6510-01-242-2271	NA	R	R	R	R	R
	Burn Treatment Water Jel, various	NA	R	R	R	R	R
	Face-shield, Industrial, Tilt: 4240-00-542-2048	NA	6	6	6	6	6
	First Aid Kit, Burn Treatment 6545-01-526-9237	1	1	1	1	1	1
	First Aid Kit, General, Ship 6545-00-116-1410	2	2	2	2	NA	2
	First Aid Kit, Small 12 unit 6545-00-922-1200	NA	NA	NA	NA	2	NA
	Gloves, Electrical: 8415-01-158-9446	NA	6	6	2	2	6
	Gloves, Unisex: 8415-00-268-7871	18	6	6	18	6	6
	Goggles, SunWindDust: 8465-01-328-8268	18	6	6	NA	6	6
	Plug, Ear (Bx): 6515-00-137-6345	1	1	1	1	NA	1
	Protector, Hearing: 4240-00-022-2946	4	6	6	4	6	6
Nomenclature and NSN							
	Goggles, Industrial, Vented: 4240-00-052-3776	6	6	6	6	6	6
	Hard Hats Orange: 8415-00-935-3136 White: 8415-00-935-3139 Brown: 8415-00-935-3135 Blue: 8415-00-935-3131	12	12	12	12	12	12
	Harness, Safety, Torso: 4240-00-022-2522	6	6	6	6	6	6
	Lanyard, Working: 4240-00-022-2518	6	6	6	6	6	6

This page intentionally left blank.

APPENDIX B

Telephonic Notification of Ground Accident

WORKSHEET FOR
TELEPHONIC NOTIFICATION OF GROUND ACCIDENT
For use of this form, see AR 385-10; the proponent agency is OCSA

Immediately notify USASC telephonically of all Class A and B accidents IAW AR 385-10, chapter 3. Phone numbers are: Commercial (334) 255-2660/2539/3410 or DSN 558-2660/2539/3410.

SHADED BLOCKS ARE FOR USASC USE ONLY	A. AGMIS CASE NUMBER	B. TIME & DATE OPS RECEIVED REPORT			
		a. Year	b. Month	c. Day	d. Time (local)

1. POINT OF CONTACT FOR ACCIDENT INFORMATION — a. Name

b. Duty: ☐ Commander ☐ Safety Officer ☐ Other (Specify) — c. Phone Number — DSN: — Commercial:

2. ACCIDENT CLASSIFICATION ☐ A ☐ B | 3. TIME & DATE OF ACCIDENT a. Year b. Month c. Day d. Time (local) | 4. PERIOD OF DAY ☐ Day ☐ Night | 5. ON/OFF DUTY ☐ On-Duty ☐ Off-Duty | 6. TYPE OF EQUIPMENT/MATERIEL INVOLVED

7. UNIT | 8. MACOM | 9. NIGHT VISION DEVISE IN USE ☐ Yes ☐ No

10. EXACT ACCIDENT LOCATION

11. ON-POST/OFF-POST? ☐ On-Post ☐ Off-Post | 12. MILITARY INSTALLATION NEAREST ACCIDENT SITE

CHECK "YES" or "NO" FOR QUESTIONS 13 THROUGH 17	Yes	No
13. EXPLOSIVE/HAZARDOUS/SENSITIVE MATERIALS INVOLVED?		
14. IF YES TO #13, ARE THEY SECURE?		
15. ACCIDENT SITE SECURED IAW AR 385-10?		
16. HAS ACCIDENT SCENE BEEN DISTURBED?		
17. IF YES TO #16, WERE PHOTOS, ETC. MADE BEFORE DISTURBING THE SCENE?		
18. WEATHER CONDITIONS		

19. PERSONNEL INVOLVED — a. No. of Personnel by Rank/Category ____ Officer ____ WO ____ Enlisted ____ Army Civilian — b. Total No. of Personnel — c. Highest Rank

20. INJURIES (Enter # of each) ____ Fatalities ____ Non-Fatal Injuries — As soon as possible, the following additional information is required on all injured personnel: name, personnel classification, degree of injury, and SSAN.

21. ACCIDENT SYNOPSIS (What happened)

22. NEWS MEDIA AWARE OF ACCIDENT ☐ Yes ☐ No | 23. NEAREST AIRFIELD — a. Nearest that can handle C-12 (4,000 ft. min.) | b. Nearest commercial airfield

24. WHO WILL INVESTIGATE? — a. Installation Level Accident Investigation (IAI) Board Appointed ☐ Yes ☐ No — b. CAI Team Dispatched ☐ Yes ☐ No — Team:

DA FORM 7306, AUG 2007 — DA FORM 7306-R, APR 94 IS OBSOLETE — APD V1.00

Figure B-1. DA Form 7306, Telephonic notification of Army accident

This page intentionally left blank.

Appendix C

Self-Contained Breathing Apparatus Specifications

- Cylinder
- Holds 4500 psi compressed air
- Fully wrapped composite construction
    - Breathing quality air (Grade D) **not oxygen**
    - Only durations of 30 and 45 minutes
    - Stored in lockers throughout ship on SCBA backpack and as spares
    - 15 year service life, requires hydrostatic testing every 3 years
    - Label contains manufacturer's name, date of manufacture, hydrostatic test information and DOT exemption number
- Valve
    - Located at neck of cylinder
    - Open and back off ¼ turn
    - Connection: CGA-347 (standard for breathing air in the pressure range of 3000-5000 psi)
    - Burst disc: actuates when pressure inside air cylinder reaches about 7200 psi
- Pressure Indicator
    - Located on valve assembly at neck of cylinder
    - Provides continuous indication of air cylinder pressure
    - Does not require calibration (shall *not* have "No Cal Required" sticker)
- Cylinder Hang Plate
    - Located on valve assembly at neck of cylinder
    - Provides mechanism for securing air cylinder to SCBA backpack
- Backpack
    - Corrosion resistant wire frame and cylinder hook (mates to cylinder hang plate)
- Cylinder Band and Latch Assembly
    - Adjustable band and latch that secures air cylinder to backpack
    - Fine adjustment can be made using vernier adjustment *while toggle lever is open* (proper adjustment = not able to turn with finger pressure when latched)
- Harness Assembly
    - Consists of two adjustable shoulder straps and an adjustable waist strap with pads
    - Waist belt has quick-release buckle and adjusters and houses holder for second stage regulator
    - Shoulder straps have pull up, push-to-release adjusters for quick adjustment
    - Flame and heat resistant Kevlar
- Remote Pressure Indicator
    - Located in front on right side shoulder strap
    - When air cylinder valve is open, provides continuous remote indication of air cylinder pressure
    - Face is fully luminescent
    - Does not require calibration (shall *not* have "No cal Required" sticker)

- Bell Alarm (Cricket Alarm)
  - Located in front on left side shoulder strap
  - When air cylinder pressure is at 23-27% of capacity, will provide a steady dinging sound to audibly alert the user of the situation
- High Pressure Hose
  - Located between cylinder valve and First-Stage regulator ("pressure reducer")
  - Delivers air at cylinder pressure to the First-Stage regulator
  - Hose is attached to air cylinder valve by the coupling nut. The high pressure seal is made by the coupling nut o-ring. Before disconnecting the coupling nut, ensure all air is bled from the high pressure hose. Air left in the hose may cause the o-ring to dislodge, resulting in inability to make a seal.
- First-Stage Regulator
  - Mounted to the left of the air cylinder
  - Reduces cylinder pressure to about 100 psi
  - Uses a redundant dual path reducing system; secondary path automatically supplies air if primary path fails
- Pressure Relief Valve
  - Located on the side of the First-Stage regulator
  - Reseatable relief valve; actuates above 185 psi
- Quick Charge Assembly
  - Located adjacent to the waist strap on the left side of the wearer
  - Air cylinder refillable without removing the air cylinder and while continuing to breathe on the SCBA

NOTE: The QC assembly should be stowed below the waist belt at all times.

- Low Pressure Hose
  - Located between pressure reducer and second stage regulator
  - Provides pressure of ~100 pounds per square inch gauge (psig) to second stage regulator
- Regulator
  - Located at the end of the low pressure hose and connects to the face piece
  - Demand regulator maintains a positive pressure in the face piece at all times
  - If face piece or seal is broken, air will flow freely from regulator
- Purge Valve
  - Red knob located on the left side of regulator (as viewed when wearing)
  - Purge valve manually overrides the Second-Stage regulator
  - Provides a constant flow of air to the face piece
  - Used for emergencies only; exit space immediately if breathing with purge valve
  - Can also be used to clear fogging in face piece
  - Rotate handle (away from wearer) to open
- Air Saver Switch
  - Black button on top of regulator
  - Stops air flow from the Second-Stage regulator
  - Press and release to actuate
  - Inhale sharply to disengage

- Removal Lever
  - Black tab on right side of regulator
  - Used to "unlock" Second-Stage regulator from face piece in order to remove it
  - To use, push tab away face and hold while turning
- "Vibralert" Alarm Assembly
  - Housed within the Second-Stage regulator
  - Alarm will sound when 20-25% of cylinder air remains
  - Alarm will also activate to indicate a problem in the First-Stage regulator
- Face Seal
  - Available in three sizes: small (green), large (black), and Xlarge (red)
  - Made of a blend of natural and synthetic rubber
- Lens
  - Single, replaceable, wide angle, clear lens
  - Made of polycarbonate with a silicone-based coating to resist abrasion and chemical attack
- Head Harness
  - Connected to the face piece by quick adjusting buckles and snap retainers
  - Made of synthetic rubber
- Voice Amplifier
  - Located in a mounting bracket over the right side voicemitter
  - Powered by one 9-volt battery

This page intentionally left blank.

GLOSSARY

SECTION I – ACRONYMS AND ABBREVIATIONS

AC	alternating current
ADE	Above deck equipment
AFFF	aqueous film forming foam
AR	Army Regulation
ARNG	Army National Guard
ASMIS	Army Safety Management Information System
b/s	breaking strength
BA	Breathing Apparatus
BD	barge derrick
BDE	Below deck equipment
BG	barge liquid (fuel)
BII	basic issue items
C	Celsius
CES	Coast earth station
CF	Causeway Ferry
CFM	Cubic feet per minute
CFR	Code of Federal Regulation
CO	carbon monoxide
CO2	carbon dioxide
COMDTINST	Commandant Instruction
COSPAS	Russian acronym for Space System for Search of Distressed Vessels
CTA	common tables of allowances
DA	Department of the Army
Db	decibel(s)
DC	damage control; direct current
DOD	Department of Defense
DODIC	Department of Defense identification code
DSN	Defense Switched Network
DTG	date-time group
E	East
EEBD	Emergency Escape Breathing Device
EOCC	Emergency Operations Control Center
EOS	Engineering Operating Station
EOW	Engineering Officer of the Watch
EPIRB	Emergency Position-Indicating Radio Beacon
ETA	estimated time of arrival
ETD	estimate time of departure
EVAC	evacuation

EXP	explosion; expiration
F	Fahrenheit
FCC	Federal Communications Commission
FED	federal
FM	field manual
FY	fiscal year
GAL	gallon(s)
GPM	gallons per minute
GPM	Ground Precautionary Message
GPS	Global Positioning System
GSA	General Services Administration
H2O	water
H2S	hydrogen sulfide
HAZMAT	hazardous material
HCF	hydro chlorofluorocarbon
HQ	headquarters
IAW	in accordance with
IDLH	immediately dangerous to life or health
JP	jet propulsion
LCM	landing craft, mechanized
LCU	landing craft, utility
LEL	lower explosive limit
LOAEL	lowest observed adverse effect level
LOTS	logistics-over-the-shore
LRG	large
LSV	logistics support vessel
MAM	Maintenance Advisory Message
MEDDAC	medical department activity
METT-TC	mission, enemy, terrain and weather, troops and support available, time available, civil considerations
MHz	megahertz
Mk	Mark
MOS	military occupational specialty
MPN	manufacture part number
MSDS	Material Safety Data Sheet
MSO	Marine Safety Office
MWT	Modular Warping Tug
N	North
NCO	noncommissioned officer
NESDIS	National Environmental Satellite, Data, and Information Service
NFPA	National Fire Protection Association
NIOSH	National Institute for Occupational Safety and Health
NO	Number
NOAA	National Oceanic and Atmospheric Administration

NSN	national stock number
OBA	oxygen breathing apparatus
OIC	officer in charge
PAM	pamphlet
PEL	permissible exposure limit
PFD	personal flotation device
PM	post meridiem
PML	personnel marker light
PN	part number
PPE	personal protection equipment
PPM	parts per million
PR	pair(s)
PSI	pounds per square inch
PSIG	pounds per square inch gauge
PVC	polyvinyl chloride
S	South
SARSAT	Search and Rescue Satellite-Aided Tracking
SART	search and rescue transponder
SC	Supply Catalog
SCBA	Self-Contained Breathing Apparatus
SES	Ship earth station
SOLAS	safety of life at sea
SOP	standing operating procedure
SOUM	Safety-Of-Use-Message
TB	technical bulletin
TBSO	Transportation Branch Safety Office
TDI	test drills and inspections
TM	technical manual
UI	unit of issue
USAR	United States Army Reserve
USCG	United States Coast Guard
VHF	very high frequency
VIP	very important person
W	With, West
WA	Washington
WI	Wisconsin
WIFCOM	wire free radio communication
WSA	Watercraft Safety Advisory
X	Extra
XLG	extra large
XLRG	extra large
XSM	extra small
XXLG	extra-extra large

SECTION II – TERMS

Actuation – Operation of a firing mechanism by an influence or series of influences in such a way that all the requirements of the mechanism for firing are met.

Ambient temperature – Outside temperature at any given altitude, preferably expressed in degrees centigrade (celsius).

Amphibious – Capable of movement on both land and water.

Anodized – Metal subjected to electrolytic action as the anode of a cell in order to coat it with a protective film.

Aqueous – Made from, by, or with water.

Backdraft - An explosion that results from combining fresh air with hot flammable fire gases, that have reached their autoignition temperatures. Large volumes of carbon monoxide and other gases can be generated by incomplete combustion in closed spaces.

Bacteriostatic – An agent that causes inhibition of the growth of bacteria without destruction.

Ballast – A heavy substance placed in such a way as to improve stability and control.

Capacitor - a device consisting of conducting plates or foils separated by thin layers of dielectric (as air or mica) with the plates on opposite sides of the dielectric layers oppositely charged by a source of voltage and the electrical energy of the charged system stored in the polarized dielectric.

Celsius – Relating to or conforming with the international thermometric scale on which the interval between the triple point of water and the boiling point of water is divided into $99.99°$ with $0.01°$ representing the triple point and $100°$ the boiling point. Also known as Centigrade. Abbreviated as C.

Collateral – Serving to support or reinforce; accompanying as secondary or subordinate.

Combustion – The usually rapid chemical process that produces heat and light.

Composite – Made up of distinct parts; relating to or being a modification of

Condensation – The conversion of a substance (as water) from the vapor state to a denser liquid or solid state usually initiated by a reduction in temperature of the vapor.

Convection – The circulatory motion that occurs in a fluid at a nonuniform temperature owing to the variation of its density and the action of gravity.

Desiccant – A drying agent (as calcium chloride).

Deviation (magnetic) – The angular difference between magnetic and compass headings.

Emission – Electromagnetic radiation from an antenna; substances discharged into the air.

Fahrenheit – Relating or conforming to a thermometric scale on which under standard atmospheric pressure the boiling point of water is at $212°$ above zero of the scale, the freezing point is at $32°$ above zero, and the zero point approximates the temperature produced by mixing equal quantities by weight of snow and common salt. Abbreviated as F.

Fire out - All visible flames have been extinguished. Smoldering fires may still be present.

Galvanized – Iron or steel coated with zinc. Immersed in molten zinc to produce a coating of zinc-iron alloy.

Grommet – An eyelet of firm material to strengthen or protect and opening or to insulate or protect something passing through it.

Hermetically sealed – Impervious to external influence. Airtight.

Hybrid – An object that has two different types of components performing essentially the same function.

Hydrostatic – Of or relating to fluids at rest or to the pressures they exert or transmit.

Hypothermia – Subnormal temperature of the body.

Integral – Essential to completeness.

Litigation – To carry on a legal contest by judicial process.

Longitudinal – Placed or running lengthwise.

Machinery space - Main and auxiliary machinery spaces which contain any of the following: installed fire fighting systems, oil fired boilers, internal combustion engines, gas turbines or steam turbines.

Neoprene – A synthetic rubber characterized by superior resistance and used especially for special-purpose clothing (as gloves and wet suits).

Out-of-Control Fire - A fire that creates untenable conditions due to heat and smoke forcing personnel to abandon the space.

Overhaul - An examination and cleanup operation. it includes finding and extinguishing hidden fire and hot embers and determining whether the fire has extended to other parts of the ship.

Oxidation – Metal combining with oxygent to form metallic oxides

Painter – A line used for securing or towing a boat. Also called a pendant.

Phenolic – A plastic made by condensation of a phenol with an aldehyde.

Polyethylene – A partially crystalline lighweight thermoplastic that is resistant to chemicals and moisture.

Polypropylene – Any of various thermoplastic plastics or fibers that are polymers of propylene.

Propagation – Ability of a transmitted radio wave to travel to a receiver by reflection off the atmospheric layers.

Pyrotechnic – A mixture of chemicals which, when ignited, is capable of reacting exothermically to produce light, heat, smoke, sound or gas.

Radiation – The transfer of radiant energy.

Spontaneous combustion – The outbreak of fire without application of heat from an external source. May occur when combustible matter is stored in bulk. It begins with a slow oxidation process under conditions not permitting ready dissipation of heat – e.g. in the center of a pile of oily rags.

Surfactant - A large group of surface acting compounds that include detergents, wetting agents and liquid soaps.

Tetrahedron – A triangle with four sides (see Figure G-1). E.g. – a pyramid.

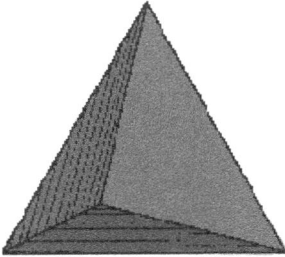

Figure G-1. Example of a Tetrahedron

Trochoid – The curve generated by a point on the radius of a circle.

Unaffected space - Any space other than the burning space.

Vapor secure - Establishing a film or foam blanket over flammable liquid to prevent vaponation. When vapors cannot reach the flames, flame production ceases and the surface is vapor secured.

References

SOURCES USED

The following lists the sources quoted or paraphrased in this publication.

AR 11-34. The Army Respiratory Protection Program. 15 February 2008

AR 15-6. Procedures for Investigating Officers and Boards of Officers. 2 October 2006

AR 56-9. Watercraft. 7 February 2002

AR 385-10. The Army Safety Program. 23 August 2007

ATSM D6193. Stitches and Seams. 14 September 1999

COMDTINST M10470.10F. Coast Guard Rescue and Survival Systems Manual. 3 January 2007

COMDTINST M16672.2D. Navigation Rules, International-Inland. 25 March 1999

CTA 50-900. Clothing and Individual Equipment. 1 September 1994

Unless otherwise indicated, DA forms are available on the APD web site (www.apd.army.mil); DD forms are available on the OSD web site (www.dtic.mil/whs/directives/infomgt/forms/formsprogram.htm)

DA Form 285. U.S. Army Accident Report

DA Form 285-AB-R. U.S. Army Abbreviated Ground Accident Report (AGAR) (LRA)

DA Form 2028. Recommended Changes to Publications and Blank Forms

DA Form 7306, Worksheet for Telephonic Notification of Ground Accident

DA Pam 385-30, Mishap Risk Management. 10 October 2007

DA Pam 385-40. Army Accident Investigation and Reporting. 1 November 1994

DD Form 314. Preventive Maintenance Schedule and Record

Publication 102. International Code of Signals for Visual, Sound, and Radio Communications, as Adopted by the Fourth Assembly of the Inter-Govermental Maritime Consultative Organization in 1965. Defense Department, National Imagery and Mapping Agency. Available online at http://www.nga.mil/portal/site/maritime/

FM 4-25.11. First Aid. 23 December 2002

FM 55-501. Marine Crewman's Handbook. 1 December 1999

TM 11-5895-1847-12&P. Operators Manual Crew Maintenance for the LCU 2000 Global Maritime Distress and Safety System (GMDSS)/LSV (1915-01-153-8801)/LCU 2000 (1905-01-154-1191)/Tug 128' Series (1925-01-247-7110). 10 July 2003

29 USC Title 29. Labor. 2 January 2006

33 USC Title 33. Navigation and Navigable Waters. 2 January 2006

46 USC Appendix Title 46. Shipping. 2 January 2006

47 USC Subchapter II. Telecommunications Carrier Compliance Payments. 2 January 2006

This page intentionally left blank.

Index

www.ingramcontent.com/pod-product-compliance
Lightning Source LLC
Chambersburg PA
CBHW081947070426
42453CB00013BA/2278